Statistics for Corpus Linguistics

EDINBURGH TEXTBOOKS IN EMPIRICAL LINGUISTICS

CORPUS LINGUISTICS
by Tony McEnery and Andrew Wilson

LANGUAGE AND COMPUTERS
A PRACTICAL INTRODUCTION TO THE COMPUTER ANALYSIS OF LANGUAGE
by Geoff Barnbrook

STATISTICS FOR CORPUS LINGUISTICS
by Michael Oakes

COMPUTER CORPUS LEXICOGRAPHY
by Vincent B. Y. Ooi

If you would like information on forthcoming titles in this series, please contact
Edinburgh University Press, 22 George Square, Edinburgh EH8 9LF

EDINBURGH TEXTBOOKS IN EMPIRICAL LINGUISTICS

Series Editors: Tony McEnery and Andrew Wilson

Statistics for Corpus Linguistics

Michael P. Oakes

EDINBURGH UNIVERSITY PRESS

© Michael P. Oakes, 1998

Edinburgh University Press
22 George Square, Edinburgh EH8 9LF

Reprinted 2003
Transferred to digital print in 2005
Typeset in 11/13pt Bembo by
Koinonia, Manchester
and printed and bound in Great Britain by
CPI Antony Rowe, Eastbourne

A CIP record for this book is available from the British Library

ISBN 0 7486 1032 4 (cased)
ISBN 0 7486 0817 6 (paperback)

Contents

Preface

I would encourage readers to get the most out of this book by using it primarily as a reference book, where relevant sections or chapters are consulted as necessary. However, if the reader does wish to read the book straight through, I hope that this will provide a good overview of all the techniques which are in use in the field. In this book, a broad interpretation of corpus-based study is made. For example, the study of stylometry is relevant to corpus linguistics even if it is not traditionally identified as such. This book is concerned more with the ways in which statistics can be applied than with precise details of how the statistics work, but space is devoted to the mathematics behind the linguistic case studies as well as to descriptions of the case studies themselves.

Acknowledgements

I would like to thank the editors, Tony McEnery and Andrew Wilson, for their support and encouragement throughout the production of this book. I would also like to thank Tony McEnery for his valuable comments on an early draft of this book.

Appendices 1 to 9, which originally appeared in Neave, H. R., *Statistics Tables for Mathematicians, Engineers, Economists and the Behavioural and Management Sciences*, George Allen and Unwin, 1978, are reprinted by permission of Routledge. Prof. Chris Butler has kindly allowed me to reproduce them in the format in which they appear in his book *Statistics in Linguistics*.

Table 2.2, which originally appeared as Table 6.1 of Church, K., Gale, W., Hanks, P. and Hindle, D., 'Using Statistics in Lexical Analysis', in Zernik, U. (ed.), *Lexical Acquisition, Exploiting On-Line Resources to Build a Lexicon*, Lawrence Erlbaum Associates, Hillsdale, New Jersey, 1992, is reprinted by permission of Lawrence Erlbaum Associates, Inc.

Figure 3.1 and Tables 3.4, 3.5 and 3.6, which originally appeared on pp. 71, 70, 58 and 72 respectively of Horvath, B. M., *Variation in Australian English*, Cambridge University Press, 1985, are reprinted by permission of Cambridge University Press.

Table 3.7, which originally appeared in Section 5.1, p. 79 of Biber, D., *Variation Across Speech and Writing*, Cambridge University Press, 1988, is reprinted by permission of Cambridge University Press.

Table 3.12, which originally appeared on p. 122 of Atwell, A. and Elliott, S. 'Dealing with Ill-Formed English Text', in Garside, R., Leech, G. and Sampson, G. (eds), *The Computational Analysis of English: A Corpus-Based Approach*, Longman, London and New York, 1987, is reprinted by permission of Addison Wesley Longman Ltd.

Table 4.5, which originally appeared as Table 9.1 on p. 129 of Renouf, A. and Sinclair, J. M., 'Collocational Frameworks in English', in Aijmer, K. and Altenberg, B. (eds), *English Corpus Linguistics*, Longman, London and New York, is reprinted by permission of Addison Wesley Longman Ltd.

Table 4.9, which originally appeared in Section 4.1 on p. 122 of Gaussier, E., Langé, J.-M. and Meunier, F., 'Towards Bilingual Terminology', 19th International Conference of the Association for Literary and Linguistic Computing, 6–9 April 1992, Christ Church Oxford, in *Proceedings of the Joint ALLC/ACH Conference,* Oxford University Press, Oxford, England, 1992, is reprinted by permission of Oxford University Press.

Table 4.11, which originally appeared as Table 1 on p. 401 of Smadja, F., 'XTRACT: an Overview', *Computers and the Humanities* 26(5–6), pp. 399–414, 1992, is reprinted with kind permission from Kluwer Academic Publishers.

Tables 4.13 and 4.14, which originally appeared as Table 1 and Figure 1 on pp. 419 and 412 respectively of Gale, W. A., Church, K. W. and Yarowsky, D., 'A Method for Disambiguating Word Senses in a Large Corpus', *Computers and the Humanities* 26(5–6), pp. 415–39, 1992, are reprinted with kind permission from Kluwer Academic Publishers.

Table 4.15, which originally appeared as Table 1 in Milton, J., 'Exploiting L1 and L2 Corpora in the Design of an Electronic Language Learning and Production Environment', in Grainger, S. (ed.), *Learner English on Computer,* Longman, Harlow, 1997, is reprinted by permission of Addison Wesley Longman Ltd.

Table 5.4, which originally appeared as Table 9.2 on p. 237 of Yule, G. U., *The Statistical Study of Literary Vocabulary,* Cambridge University Press, 1944, is reprinted by permission of Cambridge University Press.

Table 5.7, which originally appeared as Appendices C and D of Köster, P., 'Computer Stylistics: Swift and Some Contemporaries', in Wisbey, R. A. (ed.), *The Computer in Literary and Linguistic Research,* Cambridge University Press, 1971, is reprinted by permission of Cambridge University Press.

Table 5.11, which originally appeared as Table 1 on p. 154 of Leighton, J., 'Sonnets and Computers: An Experiment in Stylistic Analysis Using an Elliott 503 Computer', in Wisbey, R. A. (ed.), *The Computer in Literary and Linguistic Research,* Cambridge University Press, 1971, is reprinted by permission of Cambridge University Press.

Abbreviations

ADD	ATR Dialogue Database
ANOVA	Analysis of Variance
AP	Associated Press
ASM	Approximate String Matching
AST	Alignable Sentence Table
ATR	Advanced Telecommunications Research
BNC	British National Corpus
CALL	Computer-Assisted Language Learning
CBDF	Chi by Degrees of Freedom
COCOA	Count and Concordance on Atlas
CRATER	Corpus Resources and Terminology Extraction
DDF	Derwent Drug File
EM	Estimation-Maximisation
FA	Factor Analysis
FAG	Fager and McGowan Coefficient
FBC	French Business Correspondence
GLIM	Generalised Linear Interactive Modelling
HMM	Hidden Markov Model
IPS	Iterative Proportional Scaling
ITU	International Telecommunications Union
KUC	Kulczinsky Coefficient
KWIC	Key Words in Context
LOB	Lancaster–Oslo/Bergen
MAP	Maximum a posteriori Probability
MCC	McConnoughy Coefficient
MDS	Multi-dimensional Scaling
MFMD	Multi-Feature Multi-Dimension
MI	Mutual Information
MI2	Square Association Ratio

MI3	Cubic Association Ratio
MLE	Maximum Likelihood Estimate
MWU	Multi-Word Unit
NNS	Plural Noun
NP	Noun Phrase
OCH	Ochiai Coefficient
OCP	Oxford Concordance Program
PC	Principal Components
PCA	Principal Components Analysis
PMC	Post Modifying Clause
SAT	Sentence-Alignment Table
SEC	Spoken English Corpus
SEDM	Standard Error of Differences Between Means
SEE	Standard Error of Estimate
SES	Socio-Economic Status
SGML	Standard Generalised Mark-up Language
SMC	Simple Matching Coefficient
SPEEDCOP	Spelling Error Detection / Correction Project
SPSS	Standard Package for the Social Sciences
TACT	Text Analytic Computer Tools
VARBRUL	Variable Rule Analysis
VBZ	Third Person Singular Form of Verb
WSI	Word-Sentence Index
YUL	Yule Coefficient

Basic statistics

1 INTRODUCTION

This chapter will commence with an account of **descriptive statistics**, described by Battus (de Haan and van Hout 1986) as the useful loss of information. It also involves abstracting data. In any experimental situation we are presented with countable or measurable events such as the presence or degree of a linguistic feature in a corpus. Descriptive statistics enable one to summarise the most important properties of the observed data, such as its average or its degree of variation, so that one might, for example, identify the characteristic features of a particular author or genre. This abstracted data can then be used in **inferential statistics** (also covered in this chapter) which answers questions, formulated as hypotheses, such as whether one author or genre is different from another. A number of techniques exist for testing whether or not hypotheses are supported by the evidence in the data. Two main types of statistical tests for comparing groups of data will be described; i.e., **parametric** and **non-parametric** tests. In fact, one can never be entirely sure that the observed differences between two groups of data have not arisen by chance due to the inherent variability in the data. Thus, as we will see, one must state the **level of confidence** (typically 95 per cent) with which one can accept a given hypothesis.

Correlation and **regression** are techniques for describing the relationships in data, and are used for answering such questions as whether high values of one variable go with high values of another, or whether one can predict the value of one variable when given the value of another. **Evaluative statistics** shows how a mathematical model or theoretical distribution of data relates to reality. This chapter will describe how techniques such as regression and **loglinear analysis** enable the creation of imaginary models, and how these are compared with real-world data from direct corpus analysis. Statistical research in linguistics has traditionally been **univariate**, where the distribution of a single variable such as word frequency has been studied. However, this chapter will also cover **multivariate** techniques such as ANOVA and **multiple regression** which are concerned with the relationship between several variables. As each new statistical procedure is introduced, an example of its use in corpus linguistics will be given. A synopsis of the tests described in this chapter is given in Table 1.1.

Test	Description	Example
z score	2.4	2.4
t test for independent samples	3.2.1	3.2.2
Matched pairs t test	3.2.3	3.2.4
Wilcoxon rank sums test	3.3.1	3.3.2
Median test	3.3.3	3.3.4
Sign test	3.3.5.1	3.3.5.2
Wilcoxon matched pairs signed ranks test	3.3.5.3	3.3.5.4
Analysis of variance (ANOVA)	3.4.1	3.4.2
Chi-square test	4.1	4.2
Pearson product–moment correlation	4.3.1	4.3.2
Spearman rank correlation coefficient	4.3.3	4.3.4
Regression	4.4	4.4.1
Multiple regression	4.4.2	4.4.3
Loglinear analysis	5	5.4

Table 1.1 Statistical tests described in this chapter

This chapter will conclude with a description of **Bayesian statistics**, where we discuss our degree of belief in a hypothesis rather than its absolute probability.

2 DESCRIBING DATA

2.1 Measures of central tendency

The data for a group of items can be represented by a single score called a **measure of central tendency**. This is a single score, being the most typical score for a data set. There are three common measures of central tendency: the **mode**, the **median** and the **mean**. The mode is the most frequently obtained score in the data set. For example, a corpus might consist of sentences containing the following numbers of words: 16, 20, 15, 14, 12, 16, 13. The mode is 16, because there are more sentences with that number of words (two) than with any other number of words (all other sentence lengths occur only once). The disadvantage of using the mode is that it is easily affected by chance scores, though this is less likely to happen for large data sets.

The median is the central score of the distribution, with half of the scores being above the median and half falling below. If there is an odd number of items in the sample the median will be the central score, and if there is an even number of items in the sample, the median is the average of the two central scores. In the above example, the median is 15, because three of the sentences are longer (16, 16, 20) than this and three are shorter (12, 13, 14).

The mean is the average of all scores in a data set, found by adding up all the scores and dividing the total by the number of scores. In the sentence length example, the sum of all the words in all the sentences is 106, and the number of sentences is 7. Thus, the mean sentence length is $106/7 = 15.1$ words. The disadvantage of the mean as a measure of central tendency is that it is affected by extreme values. If the data is not **normally distributed**, with most of the items being clustered towards the lower end of the scale, for example, the median may be a more suitable measure. Normal distribution is discussed further in the section below.

2.2 Probability theory and the normal distribution
Probability theory originated from the study of games of chance, but it can be used to explain the shape of the normal distribution which is ubiquitous in nature. In a simple example, we may consider the possible outcomes of spinning an unbiased coin. The probability (p) of the coin coming up heads may be found using the formula $p = a/n$ where n is the number of equally possible outcomes, and a is the number of these that count as successes. In the case of a two-sided coin, there are only two possible outcomes, one of which counts as success, so $p = 1/2 = 0.5$. The probability of the coin coming up tails (q) is also 0.5. Since heads and tails account for all the possibilities, $p + q = 1$; and as the outcome of one spin of the coin does not affect the outcome of the next spin, the outcomes of successive spins are said to be **independent**. For a conjunction of two independent outcomes, the probability of both occurring (p) is found by multiplying the probability of the first outcome $p(a)$ by the probability of the second outcome $p(b)$. For example, the probability of being dealt a queen followed by a club from a pack of cards would be $1/13 \times 1/4 = 1/52$. In the case of a coin being spun twice, the probability of obtaining two heads would be $1/2 \times 1/2 = 1/4$, and the probability of obtaining two tails would also be $1/4$. The probability of obtaining a head followed by a tail would be $1/4$, and that of obtaining a tail followed by a head would also be $1/4$. Since there are two ways of obtaining one head and one tail (head first or tail first), the ratio of the outcomes no heads:one head:two heads is 1:2:1. Related ratios of possible outcomes for any number of trials (n) can be found using the formula $(p + q)$ to the **power** n. If n is 3, we obtain $p^3 + 3p^2q + 3pq^2 + q^3$. This shows that the ratio of times we get three heads:two heads:one head:no heads is 1:3:3:1. The number of times we expect to get r heads in n trials is called the **binomial coefficient**. This quantity, denoted by

can also be found using the formula $n!/r!\,(n-r)!$ where, for example, 4! means 4 4 x 3 x 2 x 1. The probability of success in a single trial is equal to

$$\binom{n}{r} p^r q^{n-r}$$

Plotting the probability values for all possible values of r produces the binomial probability graph.

Kenny (1982) states that any binomial distribution is completely described by p (probability) and n (number of trials). The mean is n x p, and the **standard deviation** is the square root of (n x p x q), which is usually written as $\sqrt{npq}$. The distribution is symmetrical if p and q are equally likely (**equiprobable**), as was the case for the coins, but skewed otherwise. When n is infinitely large, we obtain the normal distribution, which is a smooth curve rather than a **frequency polygon**. This curve has a characteristic bell shape, high in the centre but **asymptotically** approaching the zero axis to form a tail on either side. The normal distribution curve is found in many different spheres of life: comparing the heights of human males or the results of psychological tests, for example. In each case we have many examples of average scores, and much fewer examples of extreme scores, whether much higher or much lower than the average. This bell-shaped curve is symmetrical, and the three measures of central tendency coincide. That is to say, the mode, median and mean are equal. The normal distribution curve is shown in Figure 1.1(a). The importance of this discussion for corpus linguistics is that many of the statistical tests described in this chapter assume that the data are normally distributed. These tests should therefore only be used on a corpus where this holds true. An alternative type of distribution is the **positively skewed distribution** where most of the data is bunched below the mean, but a few data items form a tail a long way above the mean. In a **negatively skewed distribution** the converse is true; there is a long tail of items below the mean, but most items are just above it. The normal distribution has a **single modal peak**, but where the frequency curve has two peaks **bimodal distributions** also occur. Figure 1.1(b) shows a positively skewed distribution, Figure 1.1(c) shows a negatively skewed distribution and Figure 1.1(d) shows a bimodal distribution.

In corpus analysis, much of the data is skewed. This was one of Chomsky's main criticisms of corpus data, as noted by McEnery and Wilson (1996). For example, the number of letters in a word or the length of a verse in syllables are usually positively skewed. Part of the answer to criticism of corpus data based on the 'skewness' argument is that skewness can be overcome by using **lognormal distributions**. For example, we can analyse a suitably large corpus according to sentence lengths, and produce a graph showing how often each sentence length occurs in the corpus. The number of occurrences is plotted on the vertical (y) axis, and the logarithm of the sentence length in words is plotted along the horizontal (x) axis. The resulting graph will approximate to the normal distribution. Even when data is highly skewed, the normal curve is

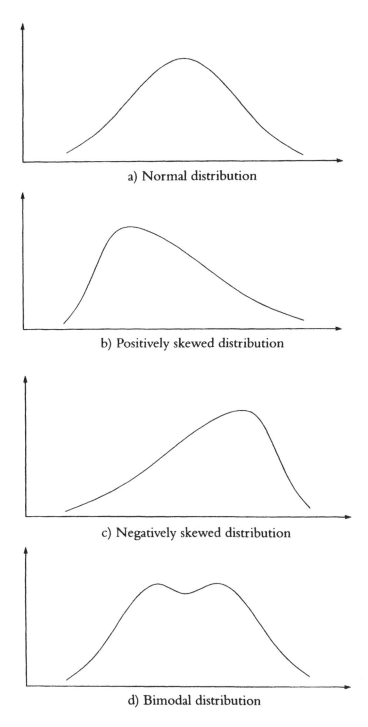

a) Normal distribution

b) Positively skewed distribution

c) Negatively skewed distribution

d) Bimodal distribution

Figure 1.1 Four distribution curves

still important. The **central limit theorem** (see Kenny 1982) states that when samples are repeatedly drawn from a population the means of the samples will be normally distributed around the population mean. This occurs whether or not the distribution of the data in the population is itself normal.

Kenny describes how, at low probabilities, particularly in the region of 0.05 which is important for determining statistical significance, the binomial distribution is no longer a good approximation to the normal distribution. Thus, we must use a different discrete distribution called the **Poisson distribution**, given by the formula

$$\text{pr}(r) = \frac{(np)^r}{r!} e^{np}$$

where e is the constant equal to about 2.72. The Poisson formula shows the probability of r events occurring where n is the number of trials and p is the probability of success at each trial. As was the case for the binomial distribution, the Poisson distribution more closely approximates to the normal distribution when n is high. The **chi-square test**, covered in Section 4.1, can be used to compare the normal and Poisson distributions with actual data.

2.3 Measures of variability

The measures of central tendency each provide us with a single value which is the most typical score for a data set. However, in describing a data set, it is also important to know the variability of scores within that data set – i.e., whether they all fall close to the most typical score, or whether the scores are spread out greatly both above and below this most typical score. The three main measures of variability are the **range**, the **variance** and the **standard deviation**. The range is simply the highest value minus the lowest value, and is thus, by definition, affected by extreme scores. To overcome this problem we may use a related measure, the **semi inter-quartile range**, which is half the difference between the value one-quarter of the way from the top end of the distribution and the value one-quarter of the way from the bottom of the distribution. For example, the mean number of characters per word in samples of text taken from nine different authors might be 5.2, 4.6, 4.4, 3.9, 3.6, 3.5, 3.4, 3.2 and 2.8. The value one-quarter of the way from the top of the distribution is 4.4, since two values are greater than this and six are less. The value one-quarter of the way from the bottom of the distribution is 3.4, since two values are less than this and six are greater. The semi inter-quartile range is (4.4–3.4)/2 = 0.5.

The variance is a measure which takes into account the distance of every data item from the mean. The simplest measure of the deviation of a single score from the mean is simply to subtract the mean value from that score. This does not translate into a suitable score of overall variability, however, since the sum of these differences over the entire data set will always be zero, since some values will be above the mean and some will be below it. To overcome this problem, the deviation of each individual score from the mean is squared, and these squares of the differences are added together to measure the total variability within the data set. By convention, this sum of squares is divided by $N-1$, where N is the number of data items, in order to obtain the variance. The greater the variance, the more scattered the data, and the less the variance, the more uniform is the data about the mean. The square root of variance is called the **standard deviation**, and may be expressed using the following formula:

$$\text{standard deviation} = \sqrt{\frac{\sum(x-\bar{x})^2}{N-1}}$$

As an example, consider the data in Table 1.2, which shows the title lengths in number of characters for 10 Indonesian short stories, gathered by the author.

Title	Length	Mean length	Deviation	Deviation squared
	x	$\bar{x}$	$x-\bar{x}$	$(x-\bar{x})^2$
1	11	13	−2	4
2	7	13	−6	36
3	7	13	−6	36
4	19	13	6	36
5	9	13	−4	16
6	11	13	−2	4
7	9	13	−4	16
8	33	13	20	400
9	9	13	−4	16
10	15	13	2	4
	Sum = 130			**Sum = 568**

Table 1.2 Calculation of the standard deviation for the title lengths of 10 stories

The variance may be found using the formula: variance $= \sum(x-\bar{x})^2/N-1$, thus variance $= 568/9 = 63.1$. The standard deviation is the square root of variance $= \sqrt{63.1} = 7.94$

2.4 The z score

The normal curve is completely defined by the mean and the standard deviation. If these are known, the shape of the entire curve can be constructed. An important property of the normal distribution curve is that if a vertical line is drawn through the normal curve at any number of standard deviations from the mean, the proportions of the area under the curve at each side of the cut-off point are always the same, as shown in Figure 1.2. A line drawn at the mean will always cut off a tail containing 50 per cent of the total curve, and a line drawn one standard deviation above the mean will always cut off a tail containing about 15.9 per cent of the area under the curve. For a line drawn two standard deviations above the mean the area in the tail will always be about 2.3 per cent of the total, and the area left by a line drawn three standard deviations above the mean will be about 0.1 per cent.

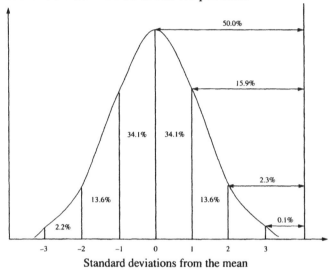

Figure 1.2 Areas of the normal distribution curve beyond a given number of standard deviations above the mean

The **z score** is a measure of how far a given value is from the mean, expressed as a number of standard deviations. The value under consideration minus the mean value is found first with z being the difference divided by the standard deviation. We might find that the mean number of times the word *there* occurs in a 1000-word sample of text written by a given author is 10, with a standard deviation of 4.2. A text sample with 8 occurrences of *there* would have a z score of $(8-10)/4.2 = -4.76$. Appendix 1 gives the proportion of the total area under the tail of the curve which lies beyond any given z value. This area is a measure of how probable an individual z score is for any test. We can discover the probability of any score in a normal distribution if we compute the z score and consult the z-score table. For example, the area beyond the z score of 1.64 is 5 per cent of the total area under the normal distribution curve. This means that the probability of encountering a value with a z score of 1.64 or more within a normally distributed data set is 5 per cent or less.

If two different data sets are both normally distributed, and the mean and standard deviation for each are known, then we can make use of the z score to compare them. In order to compare values over two tests which have different units of measurement, these values may be converted to z scores.

2.5 Hypothesis testing

The measures described so far provide means of describing data sets. However, in most cases we will not be content with merely describing data but we will need to know whether it can be used as evidence for or against experimental hypotheses. For example, Francis and Kucera (1967) found, in a study of the Corpus of Present-Day Edited American English,[1] that the mean sentence length in government documents was 25.48 words, while the mean sentence length in the corpus as a whole was 19.27. It may be that these observed differences are great enough to support the hypothesis that sentences in government documents are longer than those found in general texts, but equally, the observed differences might merely be due to the differences one would normally expect between two comparable data sets, the differences between them being purely due to chance. This second supposition is called the **null hypothesis**.

When employing statistical tests, hypotheses are stated in the null form. The null hypothesis states that there is no difference between the sample value and the population from which it was drawn. In general, the researcher hopes to disprove the null hypothesis, and show that the data item under study is significantly different from the population in general. In many fields of research, the null hypothesis is not rejected unless there are fewer than five chances in 100 of obtaining these results; in other words, that the probability of obtaining these results is less than 0.05. The purpose of statistical tests is thus to give us confidence in claims about the data, so that we may claim **statistical significance** for our results, by estimating the probability that the claims are wrong.

Where individual data items achieve scores with very low probability, i.e. they are either much higher or much lower than expected, these scores fall under the tails of the distribution curve. In a two-tailed test, we do not specify the nature of the difference (whether significantly lower or higher than the mean) in advance. In a one-tailed test, the likely direction of the difference is known beforehand. A lower z score is required to reject a one-tailed null hypothesis than to reject a two-tailed one. This is because the area under the normal distribution curve in a one-tailed test beyond the cut-off point for 0.05 probability must be divided equally between the left and right tails in a two-tail test, and thus each region has an area of 0.025. Consultation of the z-score tables shows that a z score of -1.64 leaves an area of 0.05 beyond the cut-off point. A z score of 1.96 is required to leave an area of 0.025.

2.6 Sampling

Sometimes it is not practical to study an entire corpus. For example, we may be looking at a feature which is inherently frequent, such as the definite article in English. We would absolutely not want to look at all examples retrieved from a 100,000,000-word corpus so we must select a sample which allows us to draw conclusions just as confidently as if the entire corpus had been used. A sample can be taken as representative of a population only if it is a random sample, i.e., a sample chosen in such a way that every possible sample has an equal chance of being selected. One method of sampling is called **spread sampling**, which requires the selection of small samples (taken, for example, from a corpus of text) at different starting points. The starting points of the samples are randomly chosen, perhaps by using a table of random digits. In **stratified random sampling**, often used when conducting opinion polls, the population of people to be questioned is divided according to such factors as age, sex, geography, and social and economic status.[2] These different sections of the population can be represented in the same proportion as they occur in the population as a whole, but within each section the members of the sample should be chosen by random methods.

In the context of corpus linguistics, there are three principal kinds of relationship between the sample and the population of interest. We may first wish to make generalisations about a large existing population, such as a corpus, without going to the trouble of testing it in its entirety. Secondly, we may wish to perform a study to determine whether the samples taken from a part of the corpus exhibit the characteristics of the whole corpus. In such cases we are, in fact, asking whether the relationship between the test sample and the whole corpus is that which would be expected between a random sample and the whole population. Finally, we might explore cases where the population is a hypothetical one rather than an actual one. Here we compare two existing texts to see whether their statistical behaviour is consistent with that expected for two different samples drawn from the same hypothetical corpus. Such comparisons are often made in authorship studies, for example, where an undisputed corpus of the work of one author is compared with a controversial text of doubtful authorship. A comprehensive survey of such authorship studies will be given in Chapter 5, Section 2. We need to know whether the statistics of the corpus of the undisputed work and the controversial text are such as would be expected in two samples from the same population. This population is a hypothetical one, since it consists of everything the author could have written by exhibiting the same statistical or stylistic behaviour as that found in the undisputed corpus. When performing a statistical survey of text content, we must clearly define the population from which our text samples were drawn, in order that our experiments may be reproduced. This involves, for example, recording whether or not chapter headings or stage directions are being used.

3 COMPARING GROUPS
3.1 Parametric versus non-parametric procedures
In statistics, four different types of scales of measurement should be considered, since the choice of measurement scale for the data generally determines the range of statistical tests which can be employed on the data. The concept of a **ratio scale** is exemplified by measurement in centimetres. The units on the scale are the same, and so the difference between one centimetre and two

centimetres is the same as the difference between nine centimetres and ten centimetres. An **interval scale** differs from a ratio scale in that the zero point is arbitrary – the Centigrade scale of measuring temperature, for example. Thus, although we can say that ten centimetres is twice as long as five centimetres, we cannot say that 10 degrees Centigrade is twice as hot as five degrees. If we simply record the order, rather than the distance in time, between the finishers in a cross-country race we employ an **ordinal scale**. Whenever items can be categorised – as noun, verb and adjective, for example – but the numbers we assign to the categories are arbitrary and do not reflect the primacy of any one category over the others, we have a **nominal scale**.

Parametric tests assume that dependent variables are interval- or ratio-scored. They often assume that the data is normally distributed, and that the mean and standard deviation are appropriate measures of central tendency and dispersion. However, they can work with any type of distribution with parameters. In parametric tests (such as the t test, described in Section 3.2.1), the observations should be independent – the score assigned to one case must not bias the score given to any other case.

Non-parametric tests work with frequencies and rank-ordered scales. The main advantage of non-parametric procedures is that the tests do not depend on the population being normally distributed. If sample sizes are small, a non-parametric test is best unless the population distribution is known to be normal. Non-parametric tests can treat data which can be ranked. Such data is encountered whenever a scale comparing two subjects or data items is 'less' to 'more' or 'better' to 'worse', without specifying the discrepancy exactly. Non-parametric tests such as the **chi-square test** can be used with frequency data and are typically easier to calculate by hand. Their disadvantage is that they are 'wasteful' of data. Information is lost when interval measurements are changed to ordinal ranks or nominal measurements. Parametric tests use more of the information available in the raw data since the mean and standard deviation use all the information in the data, and are thus more powerful than non-parametric tests. With a more powerful test there is less chance of making an error in rejecting or accepting the null hypothesis, and thus parametric tests are the tests of choice if all the necessary assumptions apply. Non-parametric tests can deal with interval and ratio data as well, if no distributional assumptions are to be made.

3.2 Parametric comparison of two groups
3.2.1 The t test for independent samples
Having calculated the mean and standard deviation for a group, we then want to know whether that mean is in any way exceptional, and thus need to compare the mean with that of some other group. The **t test** tests the difference between two groups for normally-distributed interval data where the mean and standard deviation are appropriate measures of central tendency and

variability of the scores. We use a t test rather than a z-score test whenever we are dealing with small samples (where either group has less than 30 items). In selecting the appropriate statistical test, we must also consider whether the data comes from two different groups (a between-groups design) or is the result of two or more measures taken from the same group (a repeated measures design). Both types of experimental design can be further subdivided into one-sample or two-sample studies.

In one-sample studies, the group mean is compared with that of the population from which it was drawn, in order to find whether the group mean is different from that of the population in general. In two-sample studies, the means from two different groups (often an experimental group and a control group) are compared to determine whether the means of these two groups differ for reasons other than pure chance.

The normal distribution is made up of individual scores. The sampling distribution of means is a distribution which, instead of being made up of individual scores, is composed of class means, and also describes a symmetric curve. The means of all groups within a population are much more similar to each other than the individual scores in a single group are to the group mean, because the mean smooths out the effect of extreme individual scores. The average of a group of means is called the population mean, μ.

A one-sample study compares a sample mean with an established population mean. The null hypothesis for a one-sample t test would be that the test scores show no difference between the sample mean and the mean of the population. To discover whether the null hypothesis is true, we must determine how far our sample mean ($\bar{x}$) is from the population mean (μ). The unit by which this difference is evaluated is the standard error of the means, which is found by dividing the standard deviation of our sample group by the square root of the sample size. Thus, the formula for the observed value of t is as follows:

$$t_{obs} = \frac{(\bar{x}) - \mu}{\text{standard error of the means}}$$

For example, an imaginary corpus containing various genres of text might contain an average of 2.5 verbs per sentence with a standard deviation of 1.2. The subset of this corpus consisting of scientific texts might have an average of 3.5 verbs per sentence with a standard deviation of 1.6. If the number of sentences in the scientific texts was 100, then the standard error of the means would be 3.5 (standard deviation of the sample group) divided by the square root of 100 (square root of the sample size) = 3.5/10 = 0.35. The observed value of t for this comparison would then be $(3.5 - 2.5)/0.35 = 2.86$.

Before we can look up this t value on the appropriate table to see if it corresponds to a statistically significant finding, we must determine the degrees of freedom in the study. The number of degrees of freedom is the number of values of the variable which are free to vary. If there are N scores contributing to a given mean value, only N–1 of them are free to vary since the other one is constrained to contribute just enough to the value of the mean. For example, if we have three variables with a mean of 10, we can select any values for the first two variables (a and b), but the third value is constrained to be 30–(a+b) if we are to maintain our mean of 10. In this simple case we therefore have two degrees of freedom. To obtain the degrees of freedom for a *t* test we use the formula N–1 where N is the number of the sample, so, in this corpus, comparison degrees of freedom = 100–1 = 99.

In the table given in Appendix 2, the probability levels are given across the top of the table, while the degrees of freedom are given in the first column. The critical value of *t* needed to reject the null hypothesis is found by checking the intersection of the row for the appropriate degrees of freedom and the selected probability cut-off point. If *t* is equal to or greater than this critical value, the null hypothesis may be rejected with confidence and we may conclude that the sample group mean is significantly greater or less than the population mean. Since there is no entry in the table for 99 degrees of freedom, we must consult the row for the next lower value of degrees of freedom, 60. The critical value of *t* for a one-tailed or directional test at the 5 per cent significance level is 1.671. Our observed value of *t* is greater than this, and thus we have disproved the null hypothesis that there is no significant difference between the number of verbs per sentence in scientific and general texts.

In making a two-sample comparison, we wish to compare the performance of two groups. The null hypothesis states that we expect any difference found between the two groups to be within what is expected due to chance for any two means in a particular population. To reject the null hypothesis, we must show that the difference falls in the extreme left or right tail of the *t* distribution. As with the one-sample study, we need to know the difference between the two means of the two groups, and to discover whether that difference is significant. To find out, we must place this difference in a sampling distribution and discover how far it is from the central point of that distribution. This time a quantity known as the **standard error of difference** between two means is used as the yardstick for our comparison rather than the standard error of a single mean. The formula for *t* in a two-sample study is as follows:

$$t = \frac{\text{difference between two sample means}}{\text{standard error of differences between means}}$$

The formula for the standard error of differences between means is

$$\text{standard error of differences between means} = \sqrt{s_e^2 / n_e + s_c^2 / n_c}.$$

where s_e = standard deviation of the experimental group, n_e = number in the experimental group, s_c = standard deviation of the control group, and n_c = number in the control group. Once again, critical values of *t* can be looked up the *t* distribution table given in Appendix 2.

When employing *t* tests, one cannot cross-compare groups. For example, one cannot compare groups A and B, A and C, then B and C. An attempt to use the *t* test for such comparisons will make it artificially easy to reject the null hypothesis. A parametric test for the comparison of three or more groups such as ANOVA (described in Section 3.4) should be used for such comparisons.

An assumption of the *t* test that relates to the normal distribution is that the data is interval-scored. It is also assumed that the mean and standard deviation are the most appropriate measures to describe the data. If the distribution is skewed, the median is a more appropriate measure of central tendency, and in such cases, a non-parametric test should be employed as described in Section 3.3.

3.2.2 Use of the t test in corpus linguistics

Imagine that the data in Table 1.3 has been obtained from a corpus of learners' English. The number of errors of a specific type occurring in each of 15 equal-length essays are shown; eight produced by students learning by traditional methods (the control group) and seven produced by students learning by a novel method that we wish to evaluate (the experimental group).

Control (n=8)	Test (n=7)
10	8
5	1
3	2
6	1
4	3
4	4
7	2
9	–

Table 1.3 Error rates produced by students taught by two different methods

The mean number of errors made by the control group was 6, with a standard deviation of 2.21. The experimental group, on the other hand, made a mean number of errors of 3, with a standard deviation of 2.27. Thus n_e was 7, s_e was 2.27, n_c was 8 and s_c was 2.21. Substituting these values into the formula for standard error of differences between means (SEDM), we obtain

$$\text{SEDM} = \sqrt{\frac{2.27 \times 2.27}{7} + \frac{2.21 \times 2.21}{8}} = \sqrt{0.736 + 0.611} = \sqrt{1.347} = 1.161$$

The difference between the two sample means was 6–3 = 3. Since t is equal to the difference between the two sample means divided by the standard error of differences between means, we obtain a t value of 3./1.161 = 2.584. The number of degrees of freedom is $n_c + n_e - 2$, or 13. Consultation of Appendix 2 shows that the critical value of significance at the 5 per cent level of t for 13 degrees of freedom is 2.16. Since the observed value of t (2.584) is greater than this, we have shown that the students learning by the novel method made significantly fewer errors in our experiment.

Church et al. (1991) provide an interesting variation on the t test. They examine the occurrence of idiomatic collocations, quantifying, for example, the relative likelihoods of encountering the phrases *powerful support* and *strong support* in their corpus. They provide the basic formula

$$t = \frac{f(\text{powerful support}) - f(\text{strong support})}{\sqrt{f(\text{powerful support}) + f(\text{strong support})}}$$

where f(powerful support) is the number of times the word pair *powerful support* appeared in the corpus. In fact, *powerful support* occurred only twice, while *strong support* occurred 175 times. After taking into account a small correction factor, this yielded a t value of N 11.94, confirming that the collocation *powerful support* was much less common than the collocation *strong support*. By employing this method, they were able to tabulate those words which are most likely to appear after *strong* (*showing, support, defence, economy, gains, winds, opposition, sales*) and those which are most likely to appear after *powerful* (*figure, minority, post, computers*).

3.2.3 The matched pairs t test

The versions of the t test described above are not appropriate for repeated-measures designs. These require another procedure called the **matched pairs t test**. This is used for other instances where correlated samples are being compared; for example, where each feature in an investigation has been observed under two different conditions, such as a word in a spoken and a written corpus, or if pairs of subjects have been matched according to any characteristic. The original t test formula must be changed slightly, because we expect that the performance of the same feature on two measures will be closer than the performance of two different features on two measures. If the scores obtained for each pair are x_1 and x_2, and the difference between them in each case is d, the formula for t in the matched pairs t test is as follows:

$$t = \frac{\sum d}{\sqrt{\dfrac{N\sum d^2 - \left(\sum d\right)^2}{N-1}}}$$

where N is the number of pairs of observations.

3.2.4 Use of the matched pairs t test in corpus linguistics

Butler (1985a) gives an example where a corpus containing the speech of 10 different speakers is analysed. The lengths of the vowels produced by those speakers was found to vary according to the consonant environment in which those vowels occurred. The recorded vowel lengths (in arbitrary time units) in two sentences, each containing one of the consonant environments under study, are shown in Table 1.4, along with their differences and the squares of their differences.

Speaker	Environment 1	Environment 2	d	d^2
1	22	26	−4	16
2	18	22	−4	16
3	26	27	−1	1
4	17	15	2	4
5	19	24	−5	25
6	23	27	−4	16
7	15	17	−2	4
8	16	20	−4	16
9	19	17	2	4
10	25	30	−5	25
			$\Sigma d = -25$	$\Sigma d^2 = 127$

Table 1.4 Lengths of a vowel in two environments

Using the formula for t for matched pairs, we obtain

$$t = \frac{-25}{\sqrt{\dfrac{(10 \times 127) - (-25)^2}{10 - 1}}} = -2.95$$

Appendix 2 gives the critical values for t with $N-1 = 9$ degrees of freedom. Ignoring the sign of t, for a one-tailed test we find that we can discount the null hypothesis that there is no difference between the vowel lengths at the 1 per cent level.

3.3 Non-parametric comparisons of two groups

Non-parametric tests are used when comparisons between two groups of data are required, but the assumptions of the t test cannot be fulfilled. One must first decide whether the comparison is between two independent groups (in which case either the median test or the Wilcoxon rank sums test could be used) or a comparison of the same group at two different time intervals (where, for example, the sign test or the Wilcoxon matched pairs signed ranks test might be used).

3.3.1 The Wilcoxon rank sums test or Mann–Whitney U test

The Wilcoxon rank sums test is also known as the Mann–Whitney U test. This test compares two groups on the basis of their ranks above and below the median and is often used when comparing ordinal rating scales rather than interval type scores. The scores for the two groups of data are first combined and ranked. The sum of the ranks found in the smaller group is then calculated, and assigned to the variable R_1, and the sum of the ranks in the larger group is assigned to the variable R_2. The test statistic U is found by use of the following formulae:

$$U_1 = N_1 N_2 + \frac{N_1(N_1 + 1)}{2} - R_1$$

$$U_2 = N_1 N_2 + \frac{N_2(N_2 + 1)}{2} - R_2$$

U is the smaller of U_1 and U_2. The distribution of U for various values of N_1 and N_2 at the 5 per cent significance level for a two-tailed test is given in Appendix 3. If the calculated value of U is smaller than or equal to the critical value we can reject the null hypothesis.

If there are approximately 20 or more subjects in each group, the distribution of U is relatively normal and the following formula may be used, where $N = N_1 + N_2$

$$z = \frac{2R_1 - N_1(N + 1)}{\sqrt{\dfrac{(N_1)(N_2)(N + 1)}{3}}}$$

For a one-tailed test at the 5 per cent significance level, the critical value of z is 1.65. If the calculated value of z is less than the critical value, we cannot reject the null hypothesis that there was no significant difference between the two groups under comparison.

3.3.2 Use of the Wilcoxon rank sums test in corpus linguistics

McEnery, Baker and Wilson (1994) describe a questionnaire which was circulated to students who used the CyberTutor corpus-based grammar-teaching computer program and to students who studied grammar in the traditional classroom manner. The questions had the format of 'How difficult/interesting/useful did you find the task?', and were answered on a five-point **Likert scale**, ranging from 'very difficult' (1 point) to 'very easy' (5 points). Aggregate scores were found for each subject in both groups. Improvised data for such a study might be as shown in Table 1.5 below. The combined scores are ranked and average scores are given to tied ranks. The sum of the ranks for the two groups (53 for the smaller group and 36 for the larger) are derived in Table 1.5.

Computer-taught	Rank	Classroom-taught	Rank
$N_2 = 7$		$N_1 = 7$	
12	2.5	10	6
9	8	7	12
14	1	8	10
11	4	6	13
8	10	10	6
12	2.5	10	6
9	8	–	–
Sum $(R_2) = 36$		**Sum $(R_1) = 53$**	

Table 1.5 Derivation of R_1 and R_2 for the Wilcoxon rank sums test

Using the formulae for U, we obtain

$$U_1 = (6 \times 7) + \frac{(6 \times 7)}{2} - 53 = 42 + 21 - 53 = 10$$

$$U_2 = (6 \times 7) + \frac{(7 \times 8)}{2} - 36 = 42 + 28 - 36 = 34$$

U is the smaller of U_2 and U_1, and thus equals 10. Consulting the critical values for U in Appendix 3, and looking up the value in the row for $N_1 = 6$ and the column for $N_2 = 7$, we find a critical value of $U = 6$. Since the value of U obtained in this experiment was 10, greater than the critical value, the null hypothesis that there was no difference between the evaluations of the human and corpus-/computer-taught groups cannot be rejected.[3]

3.3.3 The median test
This test may be employed when the data sets contain some extreme scores as the median will produce a more suitable measure of central tendency than the mean. First of all, the data sets for the two groups under investigation are pooled, and the overall median is found. If the two original groups have similar medians, then about half the observations in each of the groups would fall above the overall median and about half would fall below. To investigate the null hypothesis, a **contingency table** is constructed as shown in Table 1.6 below. A is the number in group 1 with scores above the median, B is the number in group 2 with scores above the median, C is the number in group 1 with scores below the median, and D is the number in group 2 with scores below the median.

	Group 1	Group 2
Above median	A	B
Below median	C	D

Table 1.6 Contingency table for the median test

If we employ the notation $N_1 = A + C$, $N_2 = B + D$, $N = N_1 + N_2$ and $p = (A + B)/N$, then the formula for the median test is

$$T = \frac{\dfrac{A}{N_1} - \dfrac{B}{N_2}}{\sqrt{p(1-p)\left(\dfrac{1}{N_1} + \dfrac{1}{N_2}\right)}}$$

The T value corresponds to the z values discussed previously, and so we must compare this value with the z distribution to determine whether or not we can reject the null hypothesis. If we do not predict the direction of the difference in medians beforehand, a score of 1.96 or better is required to reject the null hypothesis at the 5 per cent level.

3.3.4 Use of the median test in corpus linguistics

Hatch and Lazaraton (1991) give the following example of the use of the median test in English language testing. A test was administered to two groups of people, a group of 32 foreign students and a group of 16 immigrant students. The median score in the test (over both groups of students) was 25. It was found that 12 of the foreign students scored better than 25, and 20 less well. In the group of immigrant students, 12 people again scored over 25, but this time only four scored below this median score. These results are summarised in Table 1.7.

	Foreign	Immigrant	Total
Above median	12	12	24
Below median	20	4	24
Total	32	16	48

Table 1.7 Contingency table of language test scores for two groups of students

We first calculate p, then substitute all of our data into the formula for T:

$$p = (12 + 12)/48 = 0.5$$

$$T = \frac{(12/32) - (12/16)}{\sqrt{(.5)(1-.5)(1/32 + 1/16)}} = -2.45$$

Appendix 1 shows that T is significant at the 1 per cent level, and thus we

have disproved the null hypothesis that there was no difference in the language
test performance of the two groups.

3.3.5 Non-parametric comparisons: repeated measures

While the Mann–Whitney U test can be used as a non-parametric counterpart
of the t test for independent samples, the sign test and the Wilcoxon matched
pairs signed ranks test are the non-parametric equivalents of the matched pairs
t test.

3.3.5.1 The sign test

This test is employed when the data is measured on an ordinal scale. Matched
pairs of results are first obtained. For example, pairs of acceptability scores for a
set of sentences where it is assumed both (a) that the sentence occurs in speech
and (b) that the sentence occurs in text. For each matched pair, the sign of the
difference between result (a) and result (b) is calculated. The total number of
positive and negative differences is found, and the number of pairs with the less
frequent sign of difference is called x. N is the number of cases where the
matched pair of scores were not the same, so either a positive or a negative
difference was recorded. Appendix 4 gives critical values of x for different
values of N and different levels of significance. If the observed value of x is less
than or equal to the appropriate critical value, we can reject the null hypothesis
that the two groups of data do not differ significantly.

If N is around 25 or more, an expression derived from x becomes normally
distributed. This enables the z score to be calculated using the following formula:

$$z = \frac{N - 2x - 1}{\sqrt{N}}$$

3.3.5.2 Use of the sign test in corpus linguistics

Sentences containing honorifics retrieved from a corpus of Japanese text might
be rated on an acceptability scale from 0 to 5 according to their use in written
or spoken text. The null hypothesis would be that there is no difference in the
acceptability of Japanese sentences containing honorifics whether they occur
in written text or spoken text. The sign test would take into account whether a
sentence is rated more or less acceptable in the spoken as opposed to the written
context, but unlike the Wilcoxon matched pairs signed ranks test, does not take
into account the magnitude of the difference in acceptability. In Table 1.8 the
way the sentences were rated for acceptability according to whether they
occurred in spoken sentences or written text is shown using improvised data.

If the null hypothesis were true, we would expect the number of positive
differences to be roughly equal to the number of negative differences. In total,
we have seven positive differences and one negative difference. The number of
pairs with the less frequent sign is thus one, and this value is assigned to x. The
number of sentences (N), discounting those where no difference in
acceptability was registered, is eight. Appendix 4, which gives critical values of

Sentence	Spoken	Written	Sign of difference
1	2	4	+
2	4	5	+
3	3	3	0
4	2	4	+
5	2	5	+
6	4	2	–
7	3	3	0
8	2	4	+
9	3	5	+
10	1	2	+

Table 1.8 Acceptability ratings for Japanese sentences in speech and text

x for different values of N and different levels of significance, is then consulted. For $N=8$ at the 5 per cent significance level, the critical value of x is 1. We can thus reject the null hypothesis because our calculated value of x is less than or equal to the critical value.

3.3.5.3 The Wilcoxon matched pairs signed ranks test

This test makes the assumption that we can rank differences between corresponding pairs of observations, and thus an interval level of measurement is required. The Wilcoxon matched pairs signed ranks test is a more powerful procedure than the sign test, since we consider the degree of change as well as the direction of the differences. First of all, matched pairs of scores are obtained, and the difference between the first and second score is then found in each case. Not only is the sign of the difference recorded, but also the rank of its absolute magnitude compared with the other ranked pairs. The sum of the negative ranks and the sum of the positive ranks are both found, and the smaller of these is called W. Critical values of W according to N (the number of pairs where a difference in scores was found) and the level of significance are given in Appendix 5. If the calculated value of W is less than or equal to the critical value, the null hypothesis has been disproved.

When the number of pairs exceeds 25, the z score can be obtained from the following formula:

$$z = \frac{W - N(N+1)/4}{\sqrt{\dfrac{N(N+1)(2N+1)}{24}}}$$

3.3.5.4 A linguistic application of the Wilcoxon matched pairs signed ranks test

Butler (1985a) gives an example where this test is used to examine the number of errors made in translating two passages into French. Table 1.9 shows the raw data for this experiment as well as the differences and signed ranks of the differences for the data pairs. A mean rank is given in the case of ties, and each rank is given a sign according to whether the difference was positive or negative.

Subject no.	Passage A	Passage B	A–B	Rank
1	8	10	−2	−4.5
2	7	6	+1	+2
3	4	4	0	−
4	2	5	−3	−7.5
5	4	7	−3	−7.5
6	10	11	−1	−2
7	17	15	−2	+4.5
8	3	6	−3	−7.5
9	2	3	−1	−2
10	11	14	−3	−7.5

Table 1.9 Errors made in translating two passages into French

The pair with no difference is discounted, making the number of pairs, N, equal to 9. The sum of the positive ranks is 6.5, while the sum of the negative ranks is 38.5. The smaller of these two values, 6.5, is taken as the value of W. Critical values for W are tabulated in Appendix 5. Looking along the row for $N = 9$ and down the column for significance at the 5 per cent level, we find a critical value for W of 5. Since the calculated value of W is greater than 5, the null hypothesis holds that no significant difference between the number of errors in the two passages has been found.

3.4 Comparisons between three or more groups
3.4.1 Analysis of variance (ANOVA)
ANOVA, or **analysis of variance**, is a method of testing for significant differences between means where more than two samples are involved. ANOVA tests the null hypothesis that the samples are taken from populations having the same mean, and thus enables us to test whether observed differences between samples are greater than those arising due to chance between random samples from similar **populations**.

As described above, variance is the square of the standard deviation. ANOVA examines two sources of variance: the variance between the samples and the variance within each individual sample. If the variance between the samples is significantly greater than the variance within each group the results will suggest that the samples are not taken from the same population.

It is first of all necessary to determine the overall mean of the entire data set, found by combining all the samples under investigation. The total variability of the entire data set is then the sum of the squares of the differences between the overall mean and each data item. Next, the variation between groups must be calculated. For each sample, the difference between the overall mean and the sample mean is found and squared. The sum of these squared differences over all the samples is the between-groups variation.

The within-groups variation is then found. One method is to find the difference of each score in each sample from the mean of its own sample, then square and sum these differences. A simpler method is to simply subtract the between-groups variation from the total variation. The two types of variation are then normalised by the appropriate degrees of freedom.

The variation between groups is divided by the variation within groups, and the resulting ratio is called the F ratio. If F is close to 1, then the between-groups variance is similar to the within-groups variance and the null hypothesis holds, namely that the groups do not differ significantly from each other. If F is greater than 1, then it is more likely that the samples under test arise from different populations. Whether or not the samples differ significantly may be determined by consultation of tables of significant values of the F ratio.

3.4.2 Use of ANOVA in corpus linguistics

Kenny (1982) uses ANOVA to examine whether there is any significant difference between the number of words that three different poets (Pope, Johnson and Goldsmith) can fit into a heroic couplet. Five couplets were taken as samples of the work of each of the three poets. The overall sample of 15 couplets contained 240 words, and thus the overall mean number of words per couplet was $240/15 = 16$. The difference between the number of words in each couplet in each sample and the overall mean was found, then squared and summed across the entire data set. For example, the first line of the Pope sample contained 17 words, which differs from the overall mean by $17-16 = 1$ word. This difference squared is also 1. Performing this calculation and summing the results together for the entire data set (all three samples) produced a value of 32. This is the total variation of all the data.

To calculate the between-groups variance, the mean number of words in each sample of five couplets was found – 16 for Pope, 15 for Johnson and 17 for Goldsmith. The difference between these values and the overall mean of 16 were 0, −1 and 1 respectively. The squares of these differences were thus found to be 0, 1 and 1. These values were each weighted by multiplying the number of data items (couplets) in each sample by five, to yield weighted squared differences of 0, 5 and 5. These values were summed to yield a between-groups variance of 10. The within-groups variance was the overall variance of 32 minus the between-groups variance of 10, yielding a value of 22.

The next stage was to estimate the population variance. This was done by dividing the sums of squares by their appropriate degrees of freedom. For the sum of squares between groups, the degrees of freedom are the number of samples minus one. Thus, the estimate of population variance based on the between-groups variance is $10/2$ or 5.

The estimation of the population variance based on the variance between groups was performed in a similar manner. Each sample under test consisted of five data items, meaning that there were four degrees of freedom per group. Since there were three groups, this meant 3 x 4 = 12 degrees of freedom overall. This time the estimation of the population variance was made by dividing the variance between groups (22) by the degrees of freedom (12), yielding a value of 1.83. Dividing the between-groups estimate of population variance of 5 by the within-groups estimate of 1.83 yielded an F ratio of 2.73. Tables of the F ratio are available for various levels of significance, including the frequently accepted value of 5 per cent, and are reproduced in Appendix 6.

To consult the F-ratio table, one should find the intersection of the column corresponding to the degrees of freedom of the between-groups estimate of the population variance (number of groups minus one) and the row corresponding to the within-groups estimate (total number of data items minus the number of groups). The F value for the 5 per cent significance level found by Kenny was 3.88. Since this value was greater than the observed value of 2.73, it was concluded that the data did not show significant differences between the number of words per couplet used by each of the three poets.

4 DESCRIBING RELATIONSHIPS
4.1 The chi-square test
Nominal data are facts that can be sorted into categories such as the part of speech of each word in a corpus. They are not meant to handle subtleties of degree, but rather are measured as frequencies. We cannot state, for example, that a noun is more or less than a verb, but we can tabulate the frequency of occurrence of each part of speech encountered in the corpus, and say that nouns are more frequent than verbs. One non-parametric statistical procedure which tests the relationship between the frequencies in a display table is the **chi-square test**. This test does not allow one to make cause and effect claims, but will allow an estimation of whether the frequencies in a table differ significantly from each other. It allows the comparison of frequencies found experimentally with those expected on the basis of some theoretical model.

If we employ the null hypothesis that there is no difference between the frequencies found in each category, the first step is to decide what the frequencies would have been if there were no relationship between category and frequency. In such a case, all the frequencies would be the same, and equal to the sum of the frequencies in each cell, divided by the number of categories. This theoretical number of items per cell in the frequency table is called the expected value, E, while the actual number in each cell is called the observed value, O. For each cell, the value of $O - E$ is found and squared to give more weight to the cases where the mismatch between O and E is greatest. Finally, the value of chi-square is the sum of all the calculated values of $(O-E)^2/E$. Thus, the formula for chi-square is as follows:

$$X^2 = \sum \frac{(O-E)^2}{E}$$

When the number of degrees of freedom is 1, as in a 2 x 2 contingency table, Yates's correction factor should be applied, where 0.5 is added to each value of O if it is less than E, and 0.5 is subtracted from each value of O if it is more than E.

In a one-way design experiment, we are comparing the relation of frequencies for a single variable such as part of speech. The number of degrees of freedom is the number of cells in the frequency table minus one. In a two-way design we compare the relation of frequencies according to two different variables, such as part of speech and genre. The degrees of freedom is $(m–1)$ x $(n–1)$, if we have m levels for the first variable and n levels for the second. Expected values are found using the following formula:

$$\text{expected value} = \frac{\text{row total x column total}}{\text{grand total of items}}$$

The critical value for chi-square for a given significance level and degrees of freedom may be found in Appendix 7. If the calculated value of chi-square is greater than or equal to the critical value, we may dismiss the null hypothesis that the frequencies in the original table do not differ significantly. To use the chi-square test, the number of items investigated must be large enough to obtain an expected cell frequency of 5.

One common application of the chi-square test is to test whether two characteristics are independent, or are associated in such a way that high frequencies of one tend to coincide with high frequencies of the other. Such experiments are often based on the **contingency table**, a table in which the outcomes of an experiment are classified according to two criteria. For example, in a two-by-two contingency table, the cells are labelled A to D. In cell A we record the number of times two events occur simultaneously, for example, the number of times the words *radio* and *telescope* appear in the same sentence. In cell B we record the number of times the first event occurs but the second does not, for example, the number of times the word *radio* occurs in a sentence but *telescope* does not. In cell C we record the number of times the second event occurs but the first does not, and in cell D we record the number of instances where neither event occurs. If a is the number in cell A, b is the number in cell B and so on, and $N = a + b + c + d$, the chi-square with Yates's correction can be calculated as follows:

$$X^2 = \frac{N(|ad - bc| - N/2)^2}{(a+b)(c+d)(a+c)(b+d)}$$

To remove Yates's correction, simply replace the element $N/2$ on the top line with zero.

4.2 Use of the chi-square test in corpus linguistics

4.2.1 Two-way test to compare third person singular reference in English and Japanese texts

In Yamamoto's (1996) printed parallel corpus, the total third person singular references is 708 in the Japanese texts and 807 in the English texts. The breakdown of these figures according to five different types of third person singular reference is shown in Table 1.10.

	Japanese	English	Row total
Ellipsis	104	0	104
Central pronouns	73	314	387
Non-central pronouns	12	28	40
Names	314	291	605
Common NPS	205	174	379
Column total	708	807	1515

Table 1.10 Observed frequencies of third person singular reference in English and Japanese texts

In order to find out whether the data for the Japanese texts differs significantly from the data for English texts, a two-way chi-square test can be performed. The expected frequencies in each cell are found by first multiplying the relevant row and column totals, then dividing them by the grand total of 1515. The resulting grid of expected frequencies is given in Table 1.11.

	Japanese	English	Row total
Ellipsis	48.6	55.4	104
Central pronouns	180.9	206.1	387
Non-central pronouns	18.7	21.3	40
Names	282.7	322.3	605
Common NPS	177.1	201.9	379
Column total	708	807	1515

Table 1.11 Expected frequencies of third person singular reference in English and Japanese texts

The value of $(O-E)^2/E$ is then calculated for every cell in the grid, and the results shown in Table 1.12.

The sum of all the values shown in Table 1.12 is 258.8. Since our data is contained within five rows and two columns, we have $(5-1) \times (2-1) = 4$ degrees of freedom. Consultation of Appendix 7 at the row corresponding to four degrees of freedom and the column corresponding to a significance level

	Japanese	English
Ellipsis	63.2	55.4
Central pronouns	64.4	56.5
Non-central pronouns	2.4	2.1
Names	3.5	3.0
Common NPs	4.4	3.9

Table 1.12 $(O-E)^2/E$ for third person singular reference in English and Japanese texts

of 0.001 shows a value of 18.47. Since our calculated value of chi-square is greater than this, the breakdowns of third person singular reference types in Japanese and English have been shown to be different at the 0.001 level of significance.

4.2.2 Use of the chi-square test to show if data is normally distributed
Butler (1985a) describes a second application of the chi-square test to show whether the observed data is normally distributed. For example, the length in seconds for various vowels could be recorded. To see if this data fits the normal distribution curve closely, the data can be divided into a number (say 7) of equal time length bands (such as 0 to 0.1 seconds, 0.1 to 0.2 seconds and so on), and we can find how many vowels fall into each time band. The boundary values for each band are converted into z scores, using the formula

$$z = \frac{\text{boundary time - mean time for all the vowels}}{\text{standard deviation over all the vowels}}$$

From these z scores we wish to determine the proportion of observations expected to fall below each time boundary, which can be looked up in z-score tables, once we have determined the number of degrees of freedom. From these proportions we can first calculate the proportion of vowels expected to fall in each band for a normal distribution, and then the frequencies between successive boundaries using the formula

$$\text{expected frequency} = \text{expected proportion x sample size}$$

Butler states that the degrees of freedom in a contingency table are the number of ways in which the observed values in the individual cells of the table can vary while leaving unchanged the characteristics of the overall sample represented by the table as a whole.

In this example the distributions of observed and expected values have been made to agree on three values: the sample size, the mean and the standard deviation. These three values are not free to vary between the two distributions (observed and normal), and so we have not 7 (the number of bands) but $7-3 = 4$ degrees of freedom.

4.2.3 Use of the chi-square test with word frequency lists to measure similarity between corpora

Kilgarriff (1996b, c) discusses the possibility of using the chi-square test to examine the similarity of different text corpora. In a hypothetical example, he describes how one might find the frequencies of the common words *the*, *of*, *and* and *a* as well as the combined frequencies of all remaining words in two different corpora. His data is shown in Table 1.13.

	Corpus 1	Corpus 2
Total words	1234567	1876543
the	80123	121045
of	36356	56101
and	25143	37731
a	19976	29164

Table 1.13 Word frequency data for two different corpora

The chi-square statistic, with expected values based on probabilities in the joint corpus, is calculated as shown in Table 1.14.

	O_1	O_2	E_1	E_2	$(O_1 - E_1)^2/E_1$	$(O_2 - E_2)^2/E_2$
the	80123	121045	79828.5	121339.5	1.09	0.71
of	36356	56101	36689.3	55767.7	3.03	1.99
and	25143	37731	24950.0	37924.0	1.49	0.98
a	19976	29164	19500.0	29640.0	11.62	7.64
Remainders	1072969	1632502	1073599.2	1631871.8	0.37	0.24

Table 1.14 Calculation of the chi-square for a comparison of word frequencies in two different corpora

The sum of the items in the last two columns is 29.17, and since a 5 x 2 contingency table was used there were (5–1) x (2–1) = 4 degrees of freedom. Consultation of the chi-square distribution table shows that the critical value on four degrees of freedom at the 99 per cent significance level is 13.3, so that in this case the null hypothesis that both corpora comprise words randomly drawn from the same population may be rejected.

However, one problem with the chi-square test is that when the sample size is increased, the null hypothesis is more easily rejected. Kilgarriff shows this empirically using the words in Table 1.15 as examples. All the words in the corpus were ranked according to frequency, and the ranks of the selected words are shown in the first column.

For all but purely random populations, $(O–E)^2/E$ tends to increase with frequency. However, in natural language, words are not selected at random, and

Frequency rank	Word	$(O-E)^2/E$
1	the	18.76
41	have	10.71
321	six	5.30
5121	represent	4.53
40961	chandelier	1.15

Table 1.15 Variation in $(O-E)^2/E$ with word frequency

hence corpora are not randomly generated. If we increase the sample size, we ultimately reach the point where all null hypotheses would be rejected. The chi-square test thus cannot really be used for testing the null hypothesis.

On this basis, Kilgarriff proposes a measure of corpus similarity which uses both the chi-square statistic and word frequency information for the two corpora. This measure is called **chi by degrees of freedom** (CBDF), and provides a means whereby a similarity measure based on data for more words should be directly comparable with one based on fewer words. Each chi-square value is divided by its degrees of freedom. For example, if the chi-square is derived from the data for the most common 500 words, the appropriate degrees of freedom is 500−1 = 499. The CBDF measure was employed to examine differences between the text in various newspapers, and some of the results are shown in Table 1.16. The data shows that the broadsheet newspapers form one class, and the tabloids another.

Newspaper pair	CBDF
Mirror–Independent	14.5
Mirror–Guardian	13.2
Independent–Today	12.3
Guardian–Today	12.0
Mirror–Today	5.2
Guardian–Independent	3.8

Table 1.16 Chi by degrees of freedom for the vocabulary in pairs of newspapers

4.3 Correlation

4.3.1 The Pearson product–moment correlation coefficient

The Pearson correlation coefficient allows one to establish the strength of relationships in continuous variables. If two experimental variables are plotted against each other on a graph, this is said to be a **scatter plot**. A straight line called the **regression line** can be drawn to fit the points on the graph as closely as possible. This straight line will move up from bottom left to top right if there is a positive relationship between the two variables, or down for a negative relationship. The tighter the points cluster around the straight line, the stronger the relationship between the two variables. A scatter plot with a regression line for a positive relationship between two variables is shown in Figure 1.3.

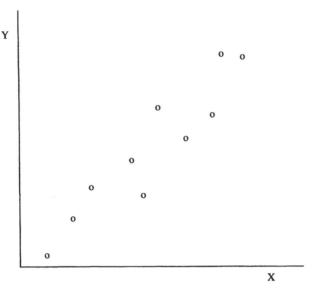

Figure 1.3 Scatter plot for two positively-related variables

The strength of the relationship between two variables can be expressed numerically using a correlation coefficient. Pearson's product–moment correlation coefficient is +1 if two variables vary together exactly. In general, a positive correlation coefficient shows that the two variables are positively correlated, where high values of the first variable are associated with high values of the second. A negative correlation where high values of the first variable are associated with low values of the second and vice versa is shown by a negative correlation coefficient. A value of −1 is obtained for a perfect negative correlation, and a value of 0 is obtained when the two variables are not correlated at all.

To calculate Pearson's correlation coefficient from raw scores, the following quantities must first be found: the sum of all values of the first variable X, the sum of all values of the second variable Y, the sum of all squares of X, the sum of all squares of Y, and the sum of the products XY over all data pairs. The correlation coefficient r is then

$$r = \frac{N\sum xy - \sum x \sum y}{\sqrt{\{N\sum x^2 - \left(\sum x\right)^2\}\{N\sum y^2 - \left(\sum y\right)^2\}}}$$

To use this formula we must assume that the scales on which X and Y are measured are truly continuous, and the scores on the two variables, X and Y are independent. The data should be normally distributed, and the relation between X and Y must be **linear**. More advanced techniques exist to describe **curvilinear relations**.

The weaker the correlation between X and Y, the greater the degree of error in regression predictions. This may be described using the quantity **standard error of estimate** (SEE), which is equal to the standard deviation of Y multiplied by the square root of $1-r^2$, where r is the correlation coefficient. r^2 is the proportion of variance in one variable accounted for by the other variable, and thus $1-r^2$ is the proportion of variance in one variable due to chance rather than the other variable.

4.3.2 Use of Pearson's product–moment correlation coefficient in corpus linguistics
Xu (1996) performed a comparison of the lengths of English words in characters with the numbers of characters in their translations into Chinese, using a bilingual English/Chinese parallel corpus. This data is shown in Table 1.17. The lengths of the English words appear in the column headed X while the lengths of the Chinese sentences appear under column Y.

S	X	Y	X^2	Y^2	XY
1	1	2	1	4	2
2	2	1	4	1	2
3	2	2	4	4	4
4	3	1	9	1	3
5	3	2	9	4	6
6	4	2	16	4	8
7	6	2	36	4	12
8	6	3	36	9	18
9	7	1	49	1	7
10	7	2	49	4	14
11	8	2	64	4	16
12	9	2	81	4	18
13	10	2	100	4	20
14	11	2	121	4	22
15	11	3	121	9	33
Total	90	29	700	61	185

Table 1.17 Correlation between the lengths of English and Chinese words

Inputting these totals into the equation, we obtain:

$$r = \frac{(15 \times 185) - (90 \times 29)}{\sqrt{\{(15 \times 700) - (90 \times 90)\}\,\{(15 \times 61) - (29 \times 29)\}}}$$

$$= \frac{2775 - 2610}{\sqrt{(10500 - 8100)(915 - 841)}}$$

$$= \frac{165}{\sqrt{2400 \times 74}} = \frac{165}{421.4} = 0.39$$

Thus, we have shown that there is a small positive correlation between the number of characters in an English word and the number of characters in its Chinese translation. We now wish to know whether there is a significant correlation between these two variables.

Appendix 8 shows the critical values of r according to the number of pairs of observations. In this study there were 15 pairs of observations, yielding a critical value of 0.441. To show significance at the 5 per cent level, the calculated value of r must be equal to or greater than the critical value. Since the calculated value of r is only 0.39, we must conclude that the correlation between the two variables under study is not significant at the 5 per cent level. A more extensive account of Xu's experiments is given by McEnery, Xu and Piao (1997).

4.3.3 Spearman's rank correlation coefficient

While Pearson's correlation coefficient requires the use of **continuous data**, Spearman's correlation coefficient may be used with **non–continuous variables**. Spearman's rank correlation coefficient is used when both variables under test are ordinal and thus may be ranked. If one variable is ordinal and the other is continuous, the continuous data may be treated as ordinal by converting it to ranked data. The formula for Spearman's rank correlation coefficient is as follows:

$$\rho = 1 - \frac{6\sum d^2}{N(N^2 - 1)}$$

4.3.4 Use of Spearman's rank correlation coefficient in corpus linguistics

McEnery, Baker and Wilson (1994) performed various experiments with the CyberTutor system, which uses annotated corpora as the basis for computer-aided tuition in assigning parts of speech to words. In one of these experiments, they wished to determine the correlation between the number of words assigned a part of speech in a set time by each of their subjects and the percentage correct. Table 1.18 shows a small unrepresentative sample of data which was obtained for the experiment. In column X we have the number of words assigned a part of speech by each subject in two hours, and in column Y we have the percentage correct. The values in columns X and Y are ranked, where the highest value is given a rank of 1, the next highest value is given a rank of 2 and so on. The resulting ranks are shown in columns X' and Y' respectively. In column d, the difference between the ranks in columns X' and Y' is shown, and the squares of these values are shown in column d^2.

S	X	Y	X'	Y'	d	d²
1	894	80.2	2	5	3	9
2	1190	86.9	1	2	1	1
3	350	75.7	6	6	0	0
4	690	80.8	4	4	0	0
5	826	84.5	3	3	0	0
6	449	89.3	5	1	4	16
						Sum = 26

Table 1.18 Correlation between speed and accuracy in a part of speech assignment test

When the values in column d^2 are all added together, the total is 26, which is the Σd^2 value required by the formula for Spearman's coefficient of correlation. The number of subjects, N, is 6. Spearman's correlation coefficient, ρ, is calculated as follows:

$$\rho = 1 - \frac{6 \times 26}{6(6^2 - 1)} = 1 - \frac{156}{210} = 1 - 0.7$$

Appendix 9 shows the critical values of the Spearman rank correlation coefficient according to the number of pairs of observations. For six pairs of observations at the 5 per cent significance level, the critical value is 0.829. The value of ρ obtained in this experiment was only 0.26, less than the critical value. So, although the two factors under investigation were shown to be positively correlated, this correlation was not found to be significant at the 5 per cent level.[4]

4.4 Regression

In most quantitative investigations, there is one variable called the **dependent variable** that we want to explain, and another variable that we believe affects the first called the **independent variable**. In an experiment, the conditions that are varied by us are the independent variables, while we measure the response of the dependent variables. **Regression** is a way of predicting the behaviour of the dependent variable according to the values of one or more independent variables. In simple regression, we predict scores on one dependent variable on the basis of scores in a single independent variable. Regression and correlation are related, so, for example, the closer the correlation is to plus or minus 1, the more accurate regression will be.

In order to predict the value of the dependent variable using regression, it is necessary to know:

1. the mean value of the independent variable scores $(\bar{X})$
2. the mean value of the scores for the dependent variable $(\bar{Y})$
3. the test value of the independent variable (X)
4. the slope of the straight line which would pass most closely to the points on a graph where the two variables are plotted against each other (b).

This straight line which best approximates to the data is called the **regression line**, and is the line that results in the smallest mean of the sum of squared vertical differences between the line and the data points.

The correlation coefficient is equal to the slope of the line of best fit in a z-score scatter plot, where the two variables are plotted together using axes where the units are standardized z scores. One way of calculating the slope of the regression line is thus to multiply the correlation coefficient by the standard deviation of the dependent variables (Y) over the standard deviation of the independent variables (X). If these correlation coefficients are not known, the slope (b) of the regression line can still be computed using raw score data using the following formula:

$$b = \frac{N\sum XY - \sum X \sum Y}{N\sum X^2 - \left(\sum X\right)^2}$$

Once the slope b has been calculated, $\hat{Y}$ (the predicted value of Y) can be found by the formula

$$\hat{Y} = \bar{Y} + b(X - \bar{X})$$

4.4.1 Use of regression in corpus linguistics

Brainerd (1974) formed a sample corpus by selecting at random seven sections of Chinese text from the *Tao Te Ching*. Table 1.19 shows the section number, the length of that section in characters (X) and the number of different Chinese characters used in that section (Y). In order to assist with the calculation of b, the slope of the regression line, the quantities X^2 and XY are also tabulated, and the sums over the entire data set for X, Y, X^2 and XY are also given.

Section	X	Y	X²	XY
1	22	20	484	440
2	49	24	2401	1176
3	80	42	6400	3360
4	26	22	676	572
5	40	23	1600	920
6	54	26	2916	1404
7	91	55	8281	5005
Sum =	362	212	22758	12877

Table 1.19 Lengths and number of characters employed in seven sections of a Chinese text

Incorporating these values into the slope equation given in the previous section gives us the following formula:

$$b = \frac{(7 \times 12877) - (362 \times 212)}{(7 \times 22758) - (362 \times 362)} = \frac{90139 - 76744}{159306 - 131044} = \frac{13395}{28262} = 0.474$$

We can use the formula $\hat{Y} = \bar{Y} + b(X - \bar{X})$ to calculate where the regression line crosses the y-axis. If we call this point a, then the complete formula for regression will be in the form $\hat{Y} = a + bX$. In order to find point a, we must work out what $\hat{Y}$ would be if X were zero. Using Brainerd's data, the mean of X (or $\bar{X}$) is 362/7 = 51.7, and the mean of Y (or $\bar{Y}$) is 212/7 = 30.3. Thus, if X is 0, $\hat{Y} = 30.3 + 0.474 (0 - 51.7) = 5.775$. This provides the complete formula for the regression line, $\hat{Y} = 5.775 + 0.474 X$.

4.4.2 Multiple regression

In simple regression we use the value of a single independent variable to predict the value of the independent variable, while **multiple regression** is used when we want to discover how well we can predict scores on a dependent variable from those of two or more independent variables. In the process of multiple regression, we estimate how much relative weight should be assigned to each of the independent variables that may affect the performance of the dependent variable. To perform multiple regression, the variables under study should either be interval or truly continuous and they should be related linearly. At least 30 data points are required for an accurate analysis. Although the calculations for multiple regressions tend to be relatively simple in concept, they are time-consuming to do by hand, so multiple regression is generally performed using statistical packages on the computer. According to Hatch and Lazaraton (1991), analysis of variance and multiple regression are probably the two most used statistical procedures in applied linguistics research.

For simple regression, the formula for the regression line was $\hat{Y} = a + bX$. For basic multiple regression, we predict $\hat{Y}$ using two independent variables (X_1 and X_2) at once, using the formula $\hat{Y} = a + b_1X_1 + b_2X_2$. b_1 and b_2 are the regression weights, analogous to the slope in simple regression, as described in Section 4.4. When X_1 is held constant, b_2 is the change in Y for a unit change in X_2. Similarly, b_1 is the effect of X_1 on Y, when X_2 is held constant.

In order to make the magnitude of weights correspond with the actual effects of each independent variable on the dependent variable, we must first standardise all the variables used in the multiple regression, by converting them to z scores. The use of z scores means that the average score in each case is zero, and thus a can be eliminated from the equation.

In order to calculate weight b_1, the correlation coefficients between each pair of variables in the study must first be found, as described in Section 4.3. These coefficients are called r_{YX_1} (the correlation between Y and X_1), r_{YX_2} and $r_{X_1X_2}$. The formula for b_1 is

$$b_1 = \frac{r_{YX_1} - r_{YX_2}r_{X_1X_2}}{1 - r_{X_1X_2}^2}$$

and the formula for b_2 is exactly analogous.

Once the multiple regression fit has been produced, it can be evaluated with a multiple correlation coefficient represented by R. The multiple correlation is the correlation between the observed Y values and those predicted from two or more X variables. If R is 0, there is no relationship between X_1, X_2 and Y, while if R is 1, X_1 and X_2 predict Y exactly. Unlike the simple correlation coefficient r, the multiple correlation R cannot be negative. The squared multiple correlation, R^2, shows the amount of variation in the independent variable that is produced by a linear combination of the independent variables. The statistical significance of R^2, which shows how confident we can be in rejecting the null hypothesis that X_1 and X_2 have no linear effects on Y, is found using the following formula:

$$F = \frac{R^2}{\left(1 - R^2\right)} \frac{\left(N - (k+1)\right)}{k}$$

N is the number of items in the data set, and k is the number of independent variables in the model. Critical values of F are given in Appendix 6. Multiple linear regression with three independent variables is very similar to regression with two independent variables. The line of best fit using three predictors is given by

$$\hat{Y} = a + b_1 X_1 + b_2 X_2 + b_3 X_3$$

4.4.3 Use of multiple regression in corpus linguistics

In Tottie's (1991) analysis of negation in English, a corpus of written English was employed. This comprised those sections of the Lancaster–Oslo/Bergen corpus devoted to press, popular lore (including magazines), belles lettres, biography, essays, and learned and scientific writings. No fictional material was used. Tottie used the VARBRUL (Variable Rule Analysis) program which allows regression to be performed using frequency data. The occurrence negation using the word *no* (no-negation) was contrasted with the occurrence of *not* forms of negation. An example of a negated clause with co-ordinated modifiers would be 'those children who will not go on to a grammar *or* senior technical school', while a contrasted negated clause would be exemplified by 'Peter saw no elephants *but* Mary saw many'. The likelihood of encountering no-negation depended on whether the negation occurred in contrastive and/or co-ordinated structures. The relationship between the three factors was expressed using the regression equation

Probability of no-negation = 0.310 Contrastive + 0.696 Co-ordination

Thus, the presence of co-ordination is more than twice as important as the presence of contrast in the prediction of whether no-negation is preferable to not-negation.

5 LOGLINEAR MODELLING

5.1 Overview

The aim of **loglinear analysis** is the analysis of frequency data as it appears in cross tabulations or contingency tables. When working with frequency data, the chi–square test is a good technique for modelling a two–variable table. However, as soon as more variables are introduced, there are many more relationships to be considered. In a three–dimensional table there may be associations between each of the pairs of variables as well as interaction between all of them.

Interaction between two independent variables is said to occur when the **degree of association** between those two variables differs between the categories of a third. Gilbert (1993) provides a UK example that there is interaction, if the magnitude of the association between social class and home tenure among those voting Conservative is not the same as the magnitude of the association between these two variables among those voting Labour.[5] In such a case the probability of voting Conservative depends not only on the independent effects of home tenure and class position but also on the effect of these influences in combination over and above their separate influences.

The chi–square test cannot handle designs with more than two independent variables. Nor can it give information about the **interaction of variables**, show which of these variables best predicts the actual distribution, or be used to test **causal claims**. However, loglinear analysis can do all these things for frequency data. It allows us to consider how many of the independent variables and interactions affect the dependent variable.

In an example given by Hatch and Lazaraton (1991), loglinear analysis is used to examine how a nominal variable such as monolingualism/bilingualism affects the odds that a student will or will not need to take a remedial course. These odds are calculated by dividing the frequency of being in one category by the frequency of not being in that category. In order to test whether these odds, called the marginal odds, are affected by the independent variables, we calculate whether the chances of being required to take remedial classes change according to whether the test subject is monolingual or bilingual. The data in their example shows that the odds of not requiring remedial classes are 1.57 (312/198) among bilingual students and 3.14 (694/221) among monolingual students, yielding an odds ratio of (3.14/1.57) = 2.

The term **model** is used by Hatch and Lazaraton to represent a statement of expectations regarding the categorical variables and their relationships to each other. For example, the above data suggests a model where +/− bilingualism (variable A) is related to +/− remedial course requirement (variable B). Using loglinear notation, this is model $\{AB\}$.

Analysing the data to examine the effect of a new variable, socio-economic status (SES), might show that the odds depend on SES as well as bilingualism. This suggests that one might propose a model that assumes it is the interaction between SES and bilingualism that influences the odds of needing remedial instruction. This model can then be tested against one that links the need for remedial instruction with bilingualism and SES, but does not assume any interaction between them. If the odds ratio in the example above does not vary with SES, then there is no interaction in the model, otherwise we can see the effect of the interaction between SES and bilingualism.

In loglinear analysis the various competing models are compared with the observed data. The so-called saturated model contains all possible effects of the variables occurring in the analysis, and of all the possible combinations of these variables. This model always fits the observed data perfectly, but in loglinear analysis the plausibility of less extensive models involving fewer variables and interactions is also examined, to see if any of these simpler models also fit the data acceptably well. The most parsimonious model, namely the one with the fewest variables and/or fewest interactions, that successfully predicts the observed odds is the one selected.

Loglinear analysis is used to construct the hypothetical frequency table that would be obtained if only those relationships specified in a model were taken into account. The resulting table is then compared with the actual data table to see whether the model is a good one. One way of generating estimates of the expected cell frequencies of each model is to use a method known as Iterative Proportional Scaling (IPS), which can be performed using the Statistical Package for the Social Sciences (SPSS) package. This will be described in detail in the following section. The Generalised Linear Interactive Modelling (GLIM) program uses an alternative and more powerful method producing similar results. Once GLIM has produced the expected frequencies for the model, these are entered into the program to produce **effect estimates** (symbolised by p or λ) for each of the variables and their interactions.

In testing the models, some computer packages with loglinear programs produce a chi-square statistic while others print out a likelihood ratio L_2. SPSS can produce either the log-likelihood chi-square (also called G^2) or the Pearson chi-square (the ordinary chi-square). The two statistics are almost the same, especially with large samples. GLIM displays the G^2 statistic under the name **scaled deviance**. The less the L_2 or chi-square (relative to the degrees of freedom) the better the model accounts for the observed data. This contrasts with the usual interpretation of chi-square, where large values tend to denote significance, but is in accordance with the fit statistics found in factor analysis.

Since we are looking for the single model that includes all the variables and interactions required to account for the original data, there is a danger that we will select a model that is 'too good'. Such a model would include spurious relationships that in reality arise solely from normal sampling error. This is the case for the saturated model, which produces a perfect fit, and other models which produce a p value close to 1. The aim is to select a model that closely fits the data but not so closely that it may include error relationships, and thus one generally looks for a model with a p value between 0.1 and 0.3 to achieve both a good fit and a parsimonious solution.

5.2 Iterative proportional scaling

In this section, the use of iterative proportional scaling (IPS) will be described by examining the data of de Haan and van Hout (1986), reproduced in Table 1.20 below. De Haan and van Hout analysed 1826 noun phrases (NP) with post modifying clauses (PMC) from a large text corpus, and classified them according to function (subject, object or prepositional complement), position (final or non-final) and length in words (1 to 5, 6 to 9 and greater than 9). They performed a loglinear analysis of this data to discover the relationships between NP function and position and length of the PMC.

Function	1–5 words	6–9 words	>9 words
(a) **Position final**			
Subject	19	21	33
Object	173	128	142
Prepositional Complement	270	284	277
(b) **Position non-final**			
Subject	156	115	84
Object	15	8	1
Prepositional Complement	43	31	26

Table 1.20 Distribution of NB functions and clause positions in the three length classes

If there is interaction between function, position and length, then the strength of the association between function and position depends on length. The association between function and length depends on position, and the association between position and length depends on function. We may discover whether the data in Table 1.20 shows interaction by constructing a model table starting from the assumption that function, position and length were associated but did not interact. This model table which shows no interaction can be compared with the actual data in Table 1.20. If the model and data tables are similar, we shall have shown that there is indeed no interaction in the data.

Table 1.20 is three-dimensional, and thus we may construct **marginal tables**[6] not only for each variable on its own, but also marginal tables showing the relationships between pairs of variables, as shown in Table 1.21. The marginal table of function by position was found by summing the data for function by position in the original data table over all word lengths, and the other marginals were found by similar means.

	Final	Non-final
Subject	73	351
Object	443	24
Prepositional complement	831	100

(a) Marginal table of function by position

	1–5 words	**6–9 words**	**>9 words**
Subject	175	136	117
Object	188	136	143
Prepositional complement	313	315	303

(b) Marginal table of function by length

	1–5 words	**6–9 words**	**>9 words**
Final	462	433	452
Non-final	214	154	111

(c) Marginal table of position by length

Table 1.21 Marginal tables for the distribution of NP functions

A model of no-association is constructed by fixing the marginals to be the same in both data and model tables. The simplest algorithm for doing this is **iterative proportional scaling**. The value of iterative models is that they can be used even when there is no exact formula for reaching a solution as is the case here. In an iterative procedure an initial estimate of the solution is made, and the estimate is then tested to see whether it is acceptably close to the solution. If not, the estimate must be refined. The testing and refinement phases are repeated until a solution has been reached.

In the process of fitting models, an arbitrary initial estimate of the model table frequencies is made. This estimate is tested by generating one of the **marginal** tables using the initial data in the model table, and comparing it with the marginal table produced for the actual data. If these do not match, the model table frequencies are updated using a standard formula (described later in this section) to bring the values closer to the observed marginal table. As the iterative process continues, the model table is matched and updated against each of the marginal tables in turn. The first estimate is to set every cell frequency in the model table equal to one. From this first estimate, the function by position marginal is calculated, and is compared with the corresponding data marginal alongside it, as shown in Table 1.22.

The estimated function by position marginal from first guess and from data marginals are clearly quite different, so this first solution is not satisfactory and must be refined. This refinement is performed by proportionately scaling the frequencies in the model table, according to the following formula:

						Function by position marginal			
First guess at a solution						**From first guess**		**From data**	
1	1	1	1	1	1	3	3	73	351
1	1	1	1	1	1	3	3	443	24
1	1	1	1	1	1	3	3	831	100

Table 1.22 IPS on the data in Table 1.21, with labels omitted

$$[\text{second guess table frequency}] = \frac{[\text{first guess table frequency}] \times [\text{data marginal entry}]}{[\text{first guess marginal entry}]}$$

For example, the frequency of the second guess table's top right corner entry is (1 x 73)/3 = 24.3. Performing the scaling for all the table frequencies yields the entire second guess table. By an analogous process, the function by length marginal is used to yield the third guess. The third guess is compared with the remaining marginal, position by length, to yield the fourth guess, and the fourth guess is compared with the function by position marginal just like the first. In fact, the order in which the marginal tables are used has no effect on the final result. The iterative process continues until all three marginals from the solution match the marginals from the data sufficiently closely for any differences to be unimportant, which is almost always achieved after 3 or 4 cycles.

The final model table has the same marginals as the data table and is the table of frequencies that would be obtained if there were no interaction between function, position and length. If the model table matches the data table exactly then there is no interaction. If the frequencies in the two tables differ, then there must be some degree of interaction, which must be quantified using statistics related to the chi-square measure, as described in the following section.

In the example above, a model table was developed starting from the assumption that all three variables were associated, but did not interact. The most obvious model to examine next is one that involves not three but two associations, again with no interaction. For example, we could try fitting the model in which the association between length and function were omitted. This is done by performing IPS using only those marginals that correspond to the relationships included in the model to be examined, namely length by position and position by function. These marginals are fixed to be identical to those in the observed data table. If this simpler model can be shown to fit the observed data, we can then try removing the length by position relationship. In this way, we can systematically work towards the simplest model that fits the data, omitting all relationships that do not contribute to the observed data. The ideas described for a three-dimensional loglinear analysis can be extended for four and higher dimension tables. The statistical methods used in this model selection process, as used by de Haan and van Hout, are described in the following section.

5.3 Selection of the best model

Altogether 19 different models can be fitted to a three-dimensional cross-tabulation for variables A, B and C. The six most important are $A*B*C$ (interaction), $A*B + B*C + A*C$ (no interaction, pairwise association), $A*B + B*C$ (no interaction, no association between A and C), $A*B + A*C$, $B*C + A*C$ and $A + B + C$ (no association). The interaction model for a three-dimensional table always fits the data perfectly, and is called the **saturation model**. As described in the previous section, the four association models are found by IPS. The no association model table entries are found using the formula

$$\text{model cell entry} = \frac{\text{row marginal x column marginal}}{\text{grand total in data table}}$$

The marginals that can be derived from a particular marginal are known as its **lower-order relatives**, where, for example, A is a lower-order relative of $A*B$. In hierarchical models A is automatically added to the model if $A*B$ is specified whenever iterative proportional scaling is used.

The question of whether one model fits better than another is addressed by quantifying the fit of a model and comparing it with the fit of rival models using a test of significance. The indicator of fit used to assess loglinear models is a measure related to the chi-square measure called the log likelihood ratio statistic G^2 (G-square, G score or log likelihood), found by using the following formula:

$$G^2 = 2\sum x_{ij}\left(\log_e x_{ij} - \log_e m_{ij}\right)$$

where x_{ij} are the data cell frequencies, m_{ij} are the model cell frequencies, $\log_e$ represents the logarithm to the base e, and the summation is carried out over all the cells in the table. G-square has a distribution that is very similar to that of the chi-square statistic, so G-square probabilities can be looked up by consulting a table of the theoretical distribution of the chi-square. Usually the computer program that calculates the model table will also provide the required number of degrees of freedom, according to the rationale provided by Gilbert (1993, p. 73):

> The more constraints a model has to satisfy, the lower the number of degrees of freedom. For loglinear models, the constraints are those marginals that are required to be identical in the model and the data. The more marginals specified in a model, the fewer the resulting degrees of freedom. In fact, a model has degrees of freedom equal to the number of cells in the table minus the total number of degrees of freedom for each of its fitted marginals. Similarly, each marginal table has degrees of freedom equal to its number of cells less the total degrees of freedom of its marginals.

A perfectly fitting model would have yielded a significance level of 100 per cent, as is the case for the interaction or saturation model. A high value of

G-square and corresponding zero significance of the simplest models shows that one or more relationships that exist in the data have been omitted. As further relationships are added to the model the degree of fit improves, leaving us to decide which model should be adopted as the simplest one that fits the data adequately. The conventional standard quoted by Gilbert is that models with a significance level of 5 per cent or more are judged to fit well. Sometimes there is a degree of subjectivity in the best model assessment; for example, where a very simple model may achieve 5 per cent significance but a slightly more complex one may have much greater significance. De Haan and van Hout found that almost all the models they tested had very high chi-square values and hence very low probability. Their best model was the one in which all two-way interactions were present. This had a chi-square value of 6.5 for four degrees of freedom, giving a probability of 16.5 per cent, which was deemed acceptable.

5.4 Example of the use of loglinear analysis in corpus linguistics: gradience in the use of the genitive

Leech, Francis and Xu (1994) performed a loglinear analysis of the data in a computer corpus to perform an empirical analysis of non-discrete categories in semantics, and in particular to demonstrate the phenomenon of gradience. Gradience means that members of two related categories differ in degree, along a scale running from 'the typical x' to 'the typical y', rather than always being assigned entirely to one category or another. The concept of gradience allows one to express the likelihood of using one form over the other in terms of probabilities rather than citing a rigid rule to always use one form in a given context.

The study of Leech, Francis and Xu is based on the analysis of the one-million-word Lancaster–Oslo/Bergen (LOB) corpus, which is a balanced corpus of early 1960s modern written British English. To illustrate their method of investigating and measuring gradience, they considered the case of the English genitive construction (as in *the president's speech*) and compared it with the frequently synonymous *of* construction (as in the *speech of the president*) These two constructions may be distinguished by the formulae [*X's Y*] and [*the Y of X*]. The factors they were seeking in their analysis were those that determine the native speaker's choice between one form rather than the other. Grammarians had previously identified a number of critical factors, including the semantic category of *X*, the semantic relation between *X* and *Y* and the style of text in which the construction occurs, so these factors were employed as the basis of the analysis.

The original aspect of their study was that the statistical technique **logistic regression** was employed. This is a type of loglinear analysis where there is one dependent variable which is to be explained by other variables. This enabled the comparison between the various theoretical models in which the previously identified factors were either present or absent, and associated or otherwise, so that the model that best fitted the classified corpus data according to a statistical test of significance could be found.

As a result of this study, Leech, Francis and Xu were able to derive from the corpus the optimal values for the factors and their subfactors (levels on an interval scale) as well as interaction effects between them. This made it possible to determine which of the factors and which of the levels within the factors were the most important in determining the choice between the genitive and the *of* construction. It was also possible to place the factors and levels in an order of importance and to discover whether any factors or levels were redundant to the analysis.

To examine the effect of genre, they used parts of the sections in the corpus for journalistic writing, scientific and learned writing, and general fiction. The semantic categories of X they employed were human nouns, animal nouns, collective nouns, place nouns, time nouns, concrete inanimate nouns and abstract nouns. Finally, the types of semantic relation between X and Y considered were possessive, subjective, objective, origin, measure, attributive and partitive. The basis for the model was the calculation of the odds in favour of the genitive, i.e., *Prob* $[X's Y]$ / *Prob* $[the Y of X]$ for any combination of factors and subfactors.

As a result of this analysis they produced a three-dimensional matrix with each cell containing a two-part frequency count in the form *frequency-of-genitive/ total-frequency*. For example, the cell devoted to the observed proportion of the genitive in journalistic style for the possessive relation and the human category is 46/72. In order to process this data for input to the GLIM statistical package, the overall three-dimensional frequency table was prepared as a set of three two-dimensional tables, one for each genre of text (journalistic, learned or fictional), each containing the observed proportion of the genitive according to semantic category and semantic relation. The table for journalistic style is reproduced in Table 1.23.

	H	A	O	P	T	C	B	Total
Possessive	46/72	0/0	8/33	16/43	0/0	0/28	1/57	71/233
Subjective	36/50	0/0	4/13	0/8	0/0	0/7	0/28	48/106
Objective	0/13	0/2	0/6	0/2	0/0	0/25	0/54	0/102
Origin	36/48	0/0	4/6	0/0	0/0	0/0	0/0	40/54
Measure	0/0	0/0	0/0	0/0	7/19	0/0	0/0	7/19
Attributive	3/7	0/0	3/7	0/3	0/0	0/12	0/21	6/50
Partitive	2/9	0/0	3/13	0/3	0/0	0/24	0/20	5/69
Total	123/199	0/2	22/68	24/59	7/19	0/96	1/180	177/633

Table 1.23 Observed proportion of the genitive in journalistic style

GLIM was used to fit a selection of statistical models to the observed data. A backward elimination procedure was adopted, with the all two-way interaction model being fitted at the initial stage. At each subsequent stage, the least important variable or interaction was selected and removed from the model using GLIM scaled deviance (likelihood ratio statistic) as the criterion for comparing nested models. Starting from the all two-way interaction model, first the category by style interaction was removed, then the relation by style interaction, and finally the relation by category interaction. The scaled deviance obtained in each case is shown in Table 1.24, as is the difference in scaled deviance in each case compared with the previous model. A difference in scaled deviance between two models will have a chi-square distribution on their difference in degrees of freedom if the term removed is unimportant. This means that the p value in the final column will be less than 0.05 if the term removed is important. The procedure was continued until all remaining variables were significant.

Model	Deviance	df	Difference in scaled deviance from previous model	Difference in df	p value
a + b + c + d	7.94	29			
a + c + d	20.26	39	12.32	10	0.2642
a + d	35.14	51	14.88	12	0.2481
a	63.01	71	27.87	20	0.1125

Table 1.24 Analysis of deviance table

In order to assess the fit of the simplest model, the scaled deviance from this model of 63.01 on 71 degrees of freedom was compared with the chi-square distribution. The critical level of chi-square at the 5 per cent significance level is 91.67, and since 63.01 is substantially below this figure, the model was deemed to fit well. It was also clearly the most parsimonious, and hence was accepted as the final model.

We can assess the importance of each of the terms in the final model by consultation of Table 1.25. All the factors are highly significant, and are very important in predicting the proportion of genitive constructs. However, when category is excluded from the model, the scaled deviance changes by the greatest value, 361, making category the most significant term, followed by relation then style.

Term deleted	Change in deviance
Style	70.28
Relation	88.33
Category	361

Table 1.25 Effect of deleting terms from the final model

To produce the final model, the relative effects of each of the subcategories of the three main effects of style, relation and category were found using the **logit function**. If we define p_{ijk} to be the probability of obtaining the genitive construct for category i, relation j and style k, then the model can be written in the form below:

$$\log_e \left(p_{ijk} / 1 - p_{ijk}\right) = K + \text{category } i + \text{ relation } j + \text{style } k$$

Using this formula with the data in the original data matrix, estimates for the relative effects of each of the subfactors in the model were calculated. The nature of this equation gives the name to loglinear analysis: the equation yields the log odds ratio and is linear in that it is the sum of unsquared terms. More complex formulae exist for models where association and interaction occur. Estimates for the various levels of the factor category are shown in Table 1.26, showing that within the category factor, the ordering of levels in terms of the fitted probabilities of choosing a genitive in preference to an *of* construction is as follows: X is human, X refers to time, X is a place, X is an organisation, X is an animal, X is abstract, X is concrete and inanimate. Similar analyses can be performed to obtain the effect estimates for relation j and style k.

Parameter Name	Description	Estimate
K	Constant	0.33
Category(1)	Human	0
Category(2)	Animal	−1.73
Category(3)	Organisation	−1.38
Category(4)	Place	−0.87
Category(5)	Time	−0.85
Category(6)	Concrete	−13.38
Category(7)	Abstract	−5.80

Table 1.26 Estimates of parameters in the final model

The net result of this analysis was to produce conclusions that could only have been arrived at by empirical means, *requiring* the use of a corpus. It was shown that all three factors of literary style, noun category and relation are important factors in determining the choice between the genitive and the *of* construction. The order of significance of the factors is that semantic class is most significant, followed by style or text type with the relation of X to Y least significant. It was possible to obtain effect estimates for the levels of all three factors, showing, for example, that the genitive is preferred to *of* more if X is human than if X is concrete and inanimate. A similar analysis not tabulated here showed that the genitive is preferred to *of* first in fictional texts, then journalistic texts, and least in learned texts.

6 BAYESIAN STATISTICS

Each of the statistical techniques described so far in this chapter employ the notion of 'absolute probability', but when using **Bayesian statistics** we discuss the 'conditional probability' of a proposition given particular evidence. In other words, we talk about belief in a hypothesis rather than its absolute probability. This degree of belief may change with the emergence of new evidence. Bayesian statistics is popular in the field of artificial intelligence, especially in expert systems; and although it is not a statistical test as such, it has a bearing on corpus linguistics. As will be described in Chapter 5, Section 2.6.1, Mosteller and Wallace (1964; see Francis 1966) used Bayesian statistics in a celebrated study of disputed authorship.

According to Krause and Clark (1993), Bayesian probability theory may be defined using the following **axioms**: Firstly $p(h\,|\,e)$, the probability of a hypothesis given the evidence, is a continuous **monotonic function**[7] in the range 0 to 1. Secondly $p(\mathit{True}\,|\,e) = 1$, meaning that the probability of a true hypothesis is one. The axiom $p(h\,|\,e) + p(\neg h\,|\,e) = 1$ means that either the hypothesis or its negation will be true. Finally, the **equality** $p(gh\,|\,e) = p(h\,|\,ge) \times p(g\,|\,e)$ gives the probability of two hypotheses being simultaneously true, which is equal to the probability of the first hypothesis, given that the second hypothesis is true, multiplied by the probability of the second hypothesis.

From this fourth axiom we can update the belief in a hypothesis in response to the observation of evidence. The equation $p(h\,|\,e) = p(e\,|\,h) \times p(h)/p(e)$ means that the updated belief in a hypothesis h on observing evidence e is obtained by multiplying the prior belief in h, $p(h)$, by the probability $p(e\,|\,h)$ that the evidence will be observed if the hypothesis is true. $p(h\,|\,e)$ is called the **a posteriori probability**, while $p(e)$ is the **a priori probability** of the evidence. Thus, conditional probability and Bayesian updating enables us to reason from evidence to hypothesis (**abduction**) as well as from hypothesis to evidence (**deduction**). If an item of evidence influences the degree of belief in a hypothesis, we have a **causal link**. The combination of such causally linked events enables one to build entire Bayesian networks. This enables probabilistic knowledge not to be represented as entries in a large joint distribution table, but rather by a network of small clusters of semantically related propositions.

Another consequence of the fourth axiom is the **chain rule**. The probability of events A_1 to A_n all occurring (the joint probability distribution) is denoted $p(A_1, A_2, ..., A_n)$, and is equal to $p(A_n\,|\,A_{n-1}, ..., A_1) \times p(A_n\,|\,A_{n-2}, ..., A_1) \times ... \times p(A_2\,|\,A_1) \times p(A_1)$. For example, the probability of encountering three words in a sequence is equal to the probability of finding the third word given the evidence of the first two words, multiplied by the probability of encountering the second word given the evidence of the first word, multiplied by the probability of the first word. Lucke (1993) describes how the chain rule is used in conjunction with the EM **algorithm**, to be described in Chapter 2, Section 2.10.3, where empirically observed data for the parts of speech frequencies for sequences of words are re-estimated for new texts.

The chain rule formula may be simplified if we know that one of the events in A_1 to A_n has no effect on our belief in the other (i.e., the two propositions are conditionally independent). If event A is conditionally independent of event B, given evidence C, then $p(A \mid B,C) = p(A \mid C)$. One example of conditional dependence is seen in the phenomenon of 'explaining away'. If event A causes event C, and event B also causes C, then if C occurs the observation of A will weaken our belief that B occurred. For example, if rain and a water sprinkler both cause a wet lawn, and the lawn is wet, then finding that the sprinkler has been left on will weaken the belief that it has been raining. Here A (rain) and B (the water sprinkler) are marginally independent, since they do not affect each other directly, but are conditionally dependent, since the observation of one affects the degree of belief in the other. Since A and B are marginally independent, we have $p(A \mid B) = p(A)$. The converse is true when we consider a case where predisposing factor A causes disease B which has associated symptom C. Once C has been confirmed, the observation of B has no further influence on our belief in A. Thus A and B are conditionally independent given C.

Krause and Clark state that, in general, the generation of a Bayesian inference model involves two steps. First the relevant qualitative inference network must be constructed, and then the relevant prior and conditional probabilities within this structure must be elicited.

6.1 Use of Bayesian statistics in corpus linguistics

Mitkov (1996) used Bayes's theorem to examine the problem of **pronominal anaphor resolution**. One hypothesis commonly used in anaphor resolution is that the focus or centre of a sentence or clause is the prime candidate for pronominal reference. Mitkov's approach was to use the Bayesian statistics approach for tracking the centre of a sentence by repeatedly updating the current probability that a certain noun or verb phrase is or is not the centre in the light of new pieces of evidence. The form of Bayes' theorem employed was as follows:

$$P(Hk \mid A) = \frac{P(Hk) \times (A \mid Hk)}{\sum P(Hi) \times P(A \mid Hi)}$$

Mitkov allowed only two possible hypotheses for a given noun or verb phrase – either it was the centre of its sentence or phrase or it was not. These two hypotheses were denoted Hy and Hn respectively, and either one can take the place of Hk in the above formula. Empirical observation of a corpus of computer science texts enabled the development of **sublanguage** dependent rules for centre tracking. Examples of such rules were 'Prefer NPs representing a domain concept to NPs which are not domain concepts' and 'If an NP is repeated throughout the discourse section, then consider it as the most probable centre'. If any of these rules applied to the noun or verb phrase in question, this was regarded as a piece of evidence. The presence or absence of such a piece of evidence was called a symptom, and A was used to denote the presence of a particular symptom in the sentence or phrase of interest.

$P(A \mid Hy)$ was the a priori probability of symptom A being observed with a noun or verb phrase which is the centre. Conversely, $P(A \mid Hn)$ is the a priori probability of the symptom being observed with a phrase which is not the centre. In Mitkov's system, the $P(A \mid Hy)$ and $P(A \mid Hn)$ values are represented by empirically discovered weight factors associated with each rule called Py and Pn respectively. The normalising factor on the bottom line of the equation, $\Sigma P(Hi) \times P(A \mid Hi)$ is found by adding $P(Hy) \times P(A \mid Hy)$ to $P(Hn) \times P(A \mid Hn)$. The a posteriori probability $P(Hk \mid A)$ is the new estimate of the probability that the verb or noun phrase is the centre, given the old probability and some new piece of evidence. Thus, in each case, Mitkov's system is to start with the initial probability of the phrase being the centre, then to consider the symptoms in turn. For each symptom the current probability is updated, taking into account whether the sentence exhibited that symptom and the weight factors Py and Pn. Mitkov tested his system for anaphor resolution, and found that results improved when traditional linguistic approaches to centre tracking were augmented with Bayesian statistical information.

7 SUMMARY

In this chapter we have looked at ways of describing sets of observed data such as linguistic features in a corpus. These descriptive statistical techniques were the three measures of central tendency, mean, median and mode which each assign a value to the most typical member of the data set. We discussed the nature of the normal distribution, with, typically, many data items clustered around its mean value and relatively few far above or below the mean. The standard deviation was introduced as a measure of variability of a data set. The z score is a standardised score which converts a data score expressed by any units to a number of standard deviations. In the section on hypothesis testing we saw how statistical tests involve a comparison between observed data and the null hypothesis which states that any difference between data sets is due to chance variation alone.

In order to compare data groups, a variety of parametric and non-parametric tests were described. A parametric test such as the t test which is used for the comparison of two groups assumes that both groups are normally distributed, being entirely described by the mean and standard deviation, and that the data must have at least an interval level of measurement. When these assumptions do not hold, non-parametric tests are used. Two forms of the t test exist; one where two different groups are compared, and one where the same group is compared at two different time intervals. Similarly, we may distinguish non-parametric tests for the comparison of independent groups such as the Mann–Whitney U test and those used for repeated measures of the same group such as the sign test. The median test is used rather than the Mann–Whitney U test when the data sets contain extreme scores, so that the median becomes the best measure of central tendency. The sign test is used for data on an ordinal scale, while the Wilcoxon matched pairs signed ranks test is used when the data is measured on an interval scale. ANOVA, or the analysis of variance, is the test of choice when three or more groups are to be compared. The chi-square test is used for the comparison of frequency data. Kilgarriff has shown that this test should be modified when working with corpus data, since the null hypothesis is always rejected when working with high-frequency words.

The degree to which two variables vary in accordance with each other is shown by measures of correlation – Pearson's product–moment correlation for continuous data and Spearman's coefficient for ranked data. Regression shows how the value of a dependent variable is determined by the value of a single independent variable, while multiple regression shows how the value of the dependent variable is determined by more than one independent variable. This chapter then gives an account of another multivariate technique, namely loglinear analysis. Here theoretical models of the data are produced by the technique of Iterative Proportional Scaling (IPS) according to whether the variables are interacting, associated or without effect. These models are compared with actual data using measures such as the G–square statistic, so that the theoretical model that most accurately fits the data can be found. The chapter concludes with an account of an alternative approach to statistics. While all the other techniques described in this chapter are concerned with absolute probabilities, Bayesian statistics is concerned with conditional probabilities and how our belief in a hypothesis can be modified by the observation of a new piece of evidence.

8 EXERCISES

1. Use the tables given in Appendices 1 to 9 to determine whether the follow-ing values are significant at the 5 per cent level:
   a) a z score of 2.01
   b) a t score of 1.75 for five degrees of freedom, two-tailed test
   c) a U score of 18 for two groups with eight and nine members respectively, non-directional test
   d) a W score of 5 for a directional test with 10 non-tied pairs of scores
   e) a Spearman rank correlation coefficient of 0.648 for 10 pairs of observa-tions in a one-tailed test.
2. In each of the following sets of circumstances, it is not desirable to use the standard t test. In each case, state which test or tests would be preferable.
   a) There are more than two groups to compare
   b) We wish to establish a repeated measures design
   c) The data is not truly continuous, for example we employ a five-point Likert Scale

d) The distribution of data is skewed.

3. A corpus is divided into five subcorpora, each containing an equal number of words. The word *element* is found to occur 40 times in the scientific subcorpus, and 15 times in each of the other four subcorpora (novels, short stories, letters and press). Use the chi-square test to decide whether the word *element* is significantly more frequent in the scientific subcorpus.

9 FURTHER READING

An excellent introduction to statistics is provided by *Statistics in Linguistics* by Christopher Butler (1985a). This book is aimed at researchers and students of general linguistics who have no previous knowledge of statistics and only a very basic knowledge of mathematics. All the material included in this chapter is covered by Butler, except for loglinear analysis and Bayesian statistics. *The Research Manual: Design and Statistics for Applied Linguistics* by Evelyn M. Hatch and Anne Lazaraton (1991) covers basic statistics from the standpoint of project design, and uses examples taken from applied linguistics. *The Computation of Style* by Anthony Kenny (1982) is a fascinating account of the statistical study of literary style. This book covers the *t* test, the chi-square test and the use of the *F* distribution, using examples drawn from authentic literary and linguistic material. Kenny describes his book as being 'written for a mathematical ignoramus with a purely humanistic background'. *Analysing Tabular Data* by Nigel Gilbert (1993) gives a clear and detailed account of loglinear analysis. The examples used in this book are taken from the social sciences. *Representing Uncertain Knowledge* by Paul Krause and Dominic Clark (1993) has a section introducing Bayesian statistics.

NOTES

1. The Brown University Standard Corpus of Present Day American English was published in 1964. It contains over one million text words of written American English taken from texts published in 1961 and was created for use with digital computers. For further details, see Francis and Kucera (1964).
2. This type of sampling would be possible using the spoken section of the British National Corpus, which encodes such data.
3. The actual data of McEnery, Baker and Wilson, which used substantially more data than that used in the above example, did show significant differences between the questionnaire responses of the two groups.
4. When using the full data set, the correlation between the number of words tagged and the percentage correct was found to be significant at the 5 per cent level.
5. The two major political parties in the United Kingdom are the Conservative and Labour parties.
6. If the frequencies along the top row of Table 1.20 are summed, the result gives the total number of NPs with subject function in the final position, irrespective of phrase length. This total is known as a marginal, because it could be added to the table along its margin.
7. Whenever the evidence is increased, the probability of the hypothesis also increases, and whenever the evidence is decreased, the probability of the hypothesis also decreases.

Information theory

1 INTRODUCTION

The following is a brief, technical introduction to this chapter. You should not worry if you do not understand the concepts introduced briefly here as it is the purpose of this chapter to explain them.

In Section 2.1 the concept of a language model will be introduced and the origins of statistically based approaches to language modelling will be described, starting in Section 2.2 with Shannon and Weaver's mathematical theory of communication. At this point it must be noted that, in information theory, the concept of **information** differs from that normally encountered. Rather than being an expression of semantic content, information here describes a level of **uncertainty**, and is highest when all the events which could occur next are equally probable. The quantity which measures information is called **entropy** (see Section 2.4), a measure of **randomness** in nature. Concepts related to entropy are **redundancy** (see Section 2.6), which is a measure of how the length of text is increased due to the statistical and linguistic rules governing a language, and redundancy-free **optimal codes** (described in Section 2.5). An extension of information theory provides the concept of **mutual information** (MI), described in Section 2.7, which is a measure of the strength of association between two events, showing whether they are more likely to occur together or independently of each other. MI enables a sequence such as words in a corpus to be compared either with the corpus as a whole or with a parallel sequence, such as the same text in another language.

The related tasks of signal processing, speech processing and text processing, first introduced in Section 2.3, will all be described in Section 2.8 as **stochastic processes** in communication, a stochastic process being a sequence of symbols, each with its own probability of occurring at a given time interval, as exemplified by the **Markov model** (see Section 2.9). This chapter will cover several approaches to the task of probabilistic language processing, in particular the automatic assignment of part-of-speech tags to the words in a corpus. Such techniques include the use of **hidden Markov models** (described in Section 2.10) which can be evaluated using the **forward–backward algorithm** and trained by the **Baum–Welch** parameter re-estimation procedure. The task of estimating the most likely sequence of states that a hidden Markov model has gone through can be performed by the **Viterbi algorithm**, which is a form of **dynamic programming**. The final concept from information theory which will be introduced is **perplexity**, which will be described in Section 2.11 as an objective measure of language model quality.

A discussion of why statistically based linguistic models are necessary to account for the variety of linguistic observations and cognitive behaviours inherent in the production of human speech patterns will be given in Section 3. In particular, it will be argued that language is not an all-or-nothing, right-or-wrong phenomenon. It is fundamentally probabilistic.

Section 4 contains a number of case histories to illustrate the use of information theory in natural language processing. The most important of these applications are part-of-speech taggers, such as CLAWS and the Cutting tagger, which are both based on Markov models. Applications of information theory in secrecy systems, poetics and stylistics, and morpheme identification will be described. The use of mutual information will be described as a means of automatic corpus annotation, for such tasks as text segmentation at the character level in morphological analysis delimited by spaces, and the identification of idiomatic collocations. Finally, Section 5 consists of a discussion of the relationship between information, the chi-square measure and the **multinomial theorem**.

2 BASIC FORMAL CONCEPTS AND TERMS

2.1 Language models

The term **linguistic model** is defined by Edmundson (1963) as an abstract representation of a natural language phenomenon. These models require quantitative data, and are thus necessarily corpus-based. A language model is always an approximation to real language, as we will see in this chapter. Examples of statistical language models are those of Markov (1916) (stochastic prediction of sequences), Shannon (1949) (redundancy of English) and Zipf (1935) (rank frequency distribution). Language models may be either **predictive** or **explicative**. Predictive models, such as Zipf's law, set out to explain future behaviour. According to Zipf's law, the rank of a word in a word frequency list ordered by descending frequency of occurrence is inversely related to its frequency. We may predict the frequency of a word from its rank using the formula

$$frequency = k \times rank^{-\gamma}$$

where k and γ are empirically found constants. The distribution of words follows Zipf's law in Chinese, English and Hebrew, but the distribution of characters conforms less so. The difference is less pronounced in Chinese, however, since many characters are complete words (Shtrikman 1994).

Explicative models exist to explain already observed phenomena, for example Shannon's use of the theory of information to estimate the degree of redundancy in the English language. The Markov model is also closely related to information theory, but in this chapter we will see that it can have a predictive function too, predicting the grammatical parts of speech of a sequence of words. In deterministic models of language typified by rule-based approaches there is no element of chance, while in stochastic models, such as those based on information theory and Markov models, events are described in terms of their statistical behaviour, particularly probability of occurrence, as a function of time. Baayen (1992) compares three models for word frequency distributions – the **lognormal law**, the **inverse Gauss–Poisson law**, and an extended version of Zipf's law.

2.2 Shannon's theory of communication

Shannon's theory of communication is concerned with the amount of information communicated by a source that generates messages, such as a human, a telephone or a newspaper. The term information does not refer to the meaning or semantic content of a single message, but to the statistical rarity of that message. The rarer a feature, the more information it carries. Information theory deals with the processes of transmission of information through communication systems, whether technical apparatus such as the telephone and the radio, or natural language communication. According to Shannon, the components of any communication system are those shown in Figure 2.1. The term message is used to describe whatever is transmitted over such a communication system.

The information source selects a desired message out of a set of possible messages. In the case of human speech, the information source is the brain. The transmitter, whose function is to transform the message into signals that can be sent along the communication channel, is the voice mechanism producing

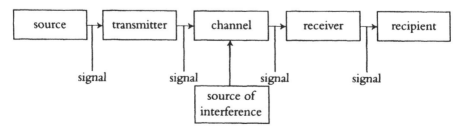

Figure 2.1 Schematic representation of a communication system

varying sound pressures. The channel itself is air. Part of the definition of the channel is its capacity, which determines such factors as the rate at which the signal is transmitted. The receiver or inverse transmitter which recreates the message from a sequence of signals is the ear and auditory nerve. The destination or recipient of the message is the brain. Noise is the term used for any form of random error arising during signal transmission, resulting in the loss of communicated data. Examples of noise are typing mistakes, distortion of waveforms and loss of bits in a computer channel. In this chapter we are mainly concerned with systems which are both discrete and noiseless. We can think of a discrete source generating the message, symbol by symbol, such as the words of written text. Successive symbols will be chosen according to certain probabilities which may depend not only on the current symbol but also on previous symbol choices. Shannon defines a physical system, or a mathematical model of a system, which produces a sequence of symbols governed by a set of probabilities as a **stochastic** process.

The application of information theory to human transmission and reception of information can be extended to include understanding in noisy conditions, where the noise is a type of undesired information. It explains how the difficulty in recognising words depends on the amount of information per word and vocabulary size, where, for example, subjects achieve better scores when selecting one of ten digits than when selecting one word from an unrestricted vocabulary. Information theory can be applied to the study of response times. For example, subjects might be requested to push on one of n buttons in response to one of n signals. The response time will depend on $\log(n)$, which is to be expected if the 'messages' in the human nervous system are optimally encoded.

2.3 Comparison between signal processing, text and speech

Information theory forces one to consider language as a code with probabilistic limitations, and permits various properties of this code to be studied using statistical methods.[1] A code is an alphabet of elementary symbols with rules for their combination. Sharman (1989) describes how natural language text, like technical codes, consists of a string of symbols (such as letters, words or multi-word groups like *put-up-with*) drawn from alphabets which are finite sets of symbols such as the set of characters from *a* to *z*, or the entire set of words in a lexicon or phrase list. A string is a finite sequence of possibly repeated symbols selected from the alphabet. In language as in technical codes, the description of the combinability of elements plays an important role. For a technical code, the rules for the combination of symbols tend to be simpler than in natural language, where there are simultaneous rules for the combinability both of meaningless units such as the phoneme or syllable, and for meaningful ones such as morphemes (Paducheva 1963).

The transmission of information is usually initiated by either a phonetic (voice) or graphic (written) source. The main difference between these is that running speech provides a continuous message, while text is a discrete source with clear divisions between words. Kruskal (1983) noted that in discrete sequences the constituent elements have been drawn from some finite alphabet such as the 20 amino acids or four nucleotides. Continuous signals are continuous functions of a variable t, typically time, where the function values may be either single numbers such as the pressure of a sound wave at time t, or a vector (whole set of values) of coefficients that describes the frequency content of a speech waveform at time t. Continuous signals such as speech are processed by converting them into discrete ones by sampling at specific time intervals. Typically 10 or 15 coordinates are used to describe the component frequencies of a sound wave at intervals of about 30 milliseconds.

Speech processing is an active research area for sequence comparison methods, where they are often referred to as **time warping** or **elastic matching**. It is difficult to match an isolated word with running speech, due to the occurrence of compression and expansion of the message when the rate of speaking increases and decreases in normal speech from instant to instant. This results in the need to expand or compress the time axis of the speech signal at various times, giving rise to the term 'time warping'. Other instances whereby the speech signal can become warped are those caused by the addition, deletion or insertion of material in everyday speech. For example, a speaker may expand the dictionary pronunciation 'offen' into the spelling pronunciation 'often' by inserting an additional t. Speech is often fragmentary, containing hesitations and corrections (insertions). In addition, spoken language contains more ellipses (deletions), inversions and situation-dependent topicalised phrasing than typical text. In a work particularly important in the context of spoken corpus data, Hosaka (1994) lists the peculiarities of speech as its tendency to include unexpected utterances other than well-formed sentences and its use of pauses as phrase demarcators. Also, Morimoto (1995) notes that speech has false starts, filled pauses and substitutions, pitch, accent and intonation. The speed and style of speech differs greatly between individuals and situations.

We should not think that text corpora are problem-free, however. The semantic content of typical textual and speech messages can also often differ. Iida (1995) states that written text such as technical documents contain mainly assertive sentences, while dialogue contains various kinds of intention expressions and makes greater use of idioms. Unlike text, speech is often accompanied by gestures. Loken-Kim et al. (1995) have shown that such deictic gestures differ in subjects speaking on the telephone and in human mediated conversation. Language is infinitely varied, whether written or spoken.

2.4 Information and entropy

Weaver (1949) states that the word 'information' in communication theory relates not so much to what you do say, as to what you could say. Thus, information is a measure of one's freedom of choice when one selects a message. The concept of information applies not only to individual messages (as the concept of meaning would) but rather to a situation as a whole. Consider a situation where the content of a message depends upon the spin of a coin. If the outcome is heads, the entire message will consist of the word 'yes', while if the outcome is tails, the message will consist of the entire text of a book. Information theory is concerned only with the fact that there are two equiprobable outcomes, and not with the fact that the semantic content of the book would be greater than that of a single word. Information theory is interested in the situation before the reception of a symbol, rather than the symbol itself. For example, information is low after encountering the letter q in English text, since there is little freedom of choice in what comes next − it is virtually always the letter u. The concept of information is primitive, like that of time, which cannot be defined either, but can be measured.

The quantity used to measure information is exactly that which is known in thermodynamics as **entropy** (H). In the physical sciences, the entropy associated with a situation is a measure of the degree of randomness (or 'shuffled-ness' if we think of a pack of cards). This is logical if we remember that information is associated with the freedom of choice we have in constructing messages. If a situation is wholly organised, not characterised by a high degree of randomness or choice, the information or entropy is low. The basic unit of information is called the **bit**, which is a contraction of the words *binary digit*. The bit is defined as the amount of information contained in the choice of one out of two equiprobable symbols such as 0 or 1, *yes* or *no*. Every message generated from an alphabet of n symbols or characters may be coded into a binary sequence. Each symbol of an n-symbol alphabet contains $\log_2 (n)$ bits of information, since that is the number of binary digits required to transmit each symbol. For example, the information in a decimal digit = $\log_2 (10) = 3.32$ bits, that of a Roman character $\log_2 (26) = 4.68$ bits and for a Cyrillic letter $\log_2 (32)$ = 5 bits.

Entropy is related to probability. To illustrate this, Weaver suggests that one should think of the situation after a message has begun with the words *in the event*. The probability that the next word is *that* is now very high, while the probability that the next word is anything else such as *elephant* is very low. Entropy is low in such situations where the probabilities are very unequal, and greatest when the probabilities of the various choices are equal. This is in accord with intuition, since minimum uncertainty occurs when one symbol is certain, and maximum uncertainty occurs when all symbols are equiprobable. The exact relationship between entropy and probabilities inherent in a system is given by the following formula:

$$H = -\left[p_1 \log_2(p_1) + p_2 \log_2(p_2) + \dots + p_n \log_2(p_n)\right]$$

The minus sign makes H positive, since the logarithms of fractions are negative. The notation $\log_2$ refers to the use of logarithms to the base 2. To convert the more commonly used logarithms to the base 10 to base 2, Shannon gives the relation

$$\log_2(M) = \log_{10}(M) / \log_{10}(2) = 3.32 \log_{10}(M)$$

In general, to change logarithms from base a to base b requires multiplication by $\log_b(a)$. Thus, in order to compute the entropy of a natural language one must:

1. count how many times each letter of the alphabet occurs
2. find the probability of occurrence of each letter by dividing its frequency by the total number of letters in the text
3. multiply each letter probability by its logarithm to the base two
4. add up all these products of probabilities and logarithms of probabilities
5. change the minus sign to a plus.

For example, the character entropy of the word *book* is calculated as follows: *b* occurs once, *o* occurs twice and *k* occurs once. Since our text consists of four letters, the probability of occurrence of *b* is $1/4 = 0.25$, that for *o* is $2/4 = 0.5$, and that for *k* is $1/4 = 0.25$. The probability of occurrence of all the other letters of the alphabet is 0, since they do not occur in the word *book*, so we need not consider them further. When we multiply each letter probability by its logarithm to the base two, for *b* we get $0.25 \times \log_2(0.25) = 0.25 \times -2 = -0.5$. For *o* we get $0.5 \times \log_2(0.5) = 0.5 \times -1 = -0.5$, and for *k* we get $0.25 \times -2 = -0.5$. Adding together these three values gives -1.5, and changing the sign gives a final entropy value of 1.5.

From the formula for entropy it follows that $H = \log_2$ (number of available choices) if all choices are equally probable. For example, $H = \log_2(2) = 1$ bit for a single spin of a two-sided coin. If all the choices are equally likely, then the more choices there are, the larger H will be. Weaver states that is logical to use a logarithmic measure, as, for example, it is natural to imagine that three binary switches can handle three times as much information as just one. One binary switch allows two different possibilities, i.e., the switch is either off or on. However, with three binary switches there are eight possibilities: off off off, off off on, off on off, off on on, on off off, on off on, on on off and on on on. $\log_2(2) = 1$ and $\log_2(8) = 3$. Hence the 'tripling of information' intuition is confirmed.

As noted in Chapter 1, Section 2.2, in systems where the probabilities of occurrence are not dependent on preceding outcomes, the probabilities are said to be independent, such as in the case of subsequent dice throws. Other systems have probabilities which do depend on previous events, such as the removal of balls from a bag which are not replaced. If we start off with three black balls and three white balls in the bag, the probability of the second ball being white will be greater if the first ball was black. In such cases we talk about conditional probabilities, since the probability of drawing a white ball is conditional on the colour of the previous ball. In a sequence of independent symbols, the amount of information for a pair of adjacent symbols is the sum of the information for each symbol.

Kahn (1966) writes that the language with the maximum possible entropy would be the one with no rules to limit it. The resulting text would be purely random, with all the letters having the same frequency, and any character equally likely to follow any other. However, the many rules of natural languages impose **structure**, and thus lower entropy. The above formula for the calculation gives the degree of entropy according to the frequency of single characters in the language, without taking into account that the probability of encountering a letter also depends on the identity of its neighbours. We can make better approximations to entropy in a natural language, by repeating the above calculation for each letter pair (**bigram**) such as *aa*, *ab* and so on, then dividing by two, because entropy is specified on a per letter basis. A better approximation still is produced by performing the above calculation for each letter triplet or **trigram** such as *aaa*, *aab* and so on, then dividing the result by three. The process of successive approximations to entropy can be repeated with ever-increasing letter group length, until we encounter long sequences of characters or *n*-grams which no longer have a valid probability of occurrence in texts. The more steps taken, the more accurate the final estimate of entropy, since each step gives a closer approximation to the entropy of language as a whole. Adhering to this process, and using a 27-letter alphabet (26 letters and a space character), Shannon (1949) found that the entropy of English was 4.03 bits per letter, for bigrams 3.32 bits per letter, and for trigrams 3.1 bits per letter. This decrease is due to the fact that each letter influences what follows it (for example, *q* is virtually always followed by *u* in English), thus imposing a degree of order. Native speakers of a language have an intuitive feel for the degree of influence that elements of a language, whether characters or words, have on the probability of what may follow. In view of this, Shannon stated that 'anyone speaking a language possesses implicitly an enormous knowledge of the statistics of a language'. Unfortunately, that knowledge, as we have noted, is vague and imprecise and only the corpus can render such accurate quantitative data.

2.5 Optimal codes

A code is an alphabet of symbols with rules for their combination. The transformation of natural language into code is called **encoding**, such as the transformation of the letter sequence of a telegram into electronic impulses, or the conversion of Chinese characters into their corresponding telegraph codes.

Consider a message written in an alphabet which has only four characters with the relative probabilities 1/2, 1/4, 1/8 and 1/8. If this is encoded using a binary code where each character is encoded by two digits (00, 01, 10 or 11), a message 1000 characters long will need 2000 binary digits to encode it. However, an alternative code could be used, where the characters are encoded 0, 10, 110, and 111. If the relative occurrence of these characters in 1000 words of text is 500, 250, 125 and 125 respectively, the number of binary digits required to encode the text is (1 x 500 + 2 x 250 + 3 x 125 + 3 x 125) = 1750. Thus, the second code makes it possible to encode text using fewer characters. The optimal code is the one which enables a message with a given quantity of information to be encoded using the fewest possible symbols. Shannon has shown that the limit to which the length of a message can be reduced when encoded in binary is determined by the quantity of information in the message. In this example, the amount of information per symbol is

$$H = -\left[\frac{1}{2}\log_2\left(\frac{1}{2}\right) + \frac{1}{4}\log_2\left(\frac{1}{4}\right) + \frac{1}{8}\log_2\left(\frac{1}{8}\right) + \frac{1}{8}\log_2\left(\frac{1}{8}\right)\right] = 1.75 \text{ bits}$$

In a message 1000 characters long, the total quantity of information is 1750 bits, so the 0, 10, 110, 111 code is optimal (Paducheva 1963).

The basic principle of optimal codes is that if the message consists of independent symbols with unequal probabilities, then assign to the most frequent symbol the shortest combination and, conversely, assign to the least frequent symbol the longest combination. If the symbols in a message are not independent, then consider code combinations for groups of symbols in the output message. Some groups may be high-frequency, others may not occur at all. To an extent, natural language is optimally encoded. Common words tend to be shorter, but English has only two one-letter words. Non-optimal codes contain an element of redundancy, which will be described in the following section.

2.6 Redundancy

The amount of information in a message increases if the number of symbols used is increased, but it decreases, for a given number of symbols, with the presence of statistical constraints in the message caused by such factors as unequal symbol probabilities and the fact that certain sequences of symbols are more likely than others. **Redundancy** is the factor by which the average lengths of messages are increased due to intersymbol statistical behaviour beyond the theoretical minimum length necessary to transmit those messages. For example, the redundancy in a message is 50 per cent if we find that we can translate it into the optimal code with the same number of symbols, and find that its length has reduced by 50 per cent (Edmundson 1963).

Having found the entropy or information per symbol of a certain information source, this can be compared to the maximum value this entropy could have, if the source were to continue to use the same symbols. The entropy is maximum when the symbols are independent and equiprobable. When each symbol carries maximum information, the code is said to be **utilised to capacity** or **optimal**. For a given number of symbols n, maximum entropy $H_{max} = \log_2(n)$. Thus Kahn (1966) calculates the maximum entropy of an alphabet of 27 symbols (26 letters and a space symbol) as $\log_2(27)$ or 4.76 bits per letter. Actual entropy H, maximum entropy H_{max}, relative entropy H_{rel} and redundancy R are related by the following formulae:

$$H_{rel} = H/H_{max}$$
$$R = 1 - H_{rel}$$

Thus, the ratio of actual entropy divided by the maximum entropy is called **relative entropy**, and 1 − relative entropy = redundancy. Both relative entropy and redundancy are expressed as percentages. If the entropy of English were taken to be one bit per letter, redundancy would be 1−(1/4.76) or about 75 per cent.

Redundancy is the fraction of the structure of the message which is determined not by the free choice of the sender, but rather by the statistical rules governing the use of the symbols in question. If the redundant part of the message were missing, the message would still be essentially complete. However, Hood-Roberts (1965) points out that redundancy is an essential property of language which permits one to understand what is said or written even when a message is corrupted by considerable amounts of noise. The presence of redundancy enables us to reconstruct missing components of messages.

As an example of sources of redundancy in language, not all elements of language such as letters, phonemes or words have the same frequency. For example, according to Hood-Roberts (1965), the relative frequency of phonemes in English varies from 11.82 per cent to 0.03 per cent. Dewey (1923) gives the relative frequencies of many features of the English language. The same effect is noted for bigrams, trigrams and other n-grams. As longer sequences are considered, the proportion of meaningful messages to the total number of possible messages decreases. A second example is that more than one letter or letter combination may encode the same sound, such as the letter sequences *ks* and *x*, *kw* and *q*, and *c* which can represent the sound of phonemes /k/ or /s/ in British English spelling; and third is the existence of uneven conditional probabilities, where, for example, the letter *q* in English is almost always followed by the letter *u*, and almost never followed by anything else. For example, uneven conditional probabilities are seen in that the word sequence *programming languages* is much more likely than the sequence *languages programming*.

As a result of redundancy, electronic text files can be compressed by about 40 per cent without any loss of information. More efficient coding of natural language can sometimes be achieved by permitting a certain deterioration in the original message, by using vowelless English, for example, or the conversion of colloquial language into terse telegraphic style. Both of these are possible due to redundancy (Bar-Hillel 1964). An example of almost vowelless English given by Denning (1982) is *mst ids cn b xprsd n fwr ltrs bt th xprnc s mst nplsnt.*

Denning describes how the conscious introduction of redundancy into a coding system enables the creation of self-correcting codes. Coding is not only used for greater efficiency in terms of time and cost, but also to improve the transmission of information. With self-correcting codes, not all possible symbol combinations are used as code combinations, according to certain rules. A distortion of one of the symbols changes a code combination into a set of symbols that is not a legitimate code combination. If the distinction among code combinations is great enough, one can not only find out that an error has occurred, but also predict rather accurately what was the original correct one. An optimal code which has no redundancy and uses the least possible number of symbols must be perfect for correct understanding.

2.7 Mutual information

Mutual information, and other co-occurrence statistics, are slowly taking a central position in corpus linguistics. As a measure, it is described in other books in this series (McEnery and Wilson 1996; Ooi 1998). My aim here is to describe mutual information in the context of established information theory. Consider h and i to be events which both occur within sequences of events. In a linguistic context, h might be a word in an English sentence while i might be a word in the equivalent French sentence; or h might be an input word to a noisy channel while i is an output word from that channel. h and i might be members of the same sequence; for example, two words which occur in an idiomatic collocation. Sharman (1989) describes how mutual information, denoted $I(h;i)$, shows us what information is provided about event h by the occurrence of event i. $P(h \mid i)$ is the probability of event h having occurred when we know that event i has occurred, called the **a posteriori probability**, and $P(h)$ is the probability of event h having occurred when we do not know whether or not event i has occurred, called the **a priori probability**. For example, $P(h \mid i)$ could be the probability that the third word in an English sentence might be *cat,* given that the fourth word in the equivalent French sentence is *chat*, while $P(h)$ could be the probability of the word *cat* occurring in an English sentence regardless of what words appear in its French translation. The relation between $I(h;i)$, the a posteriori probability of h and the a priori probability of h is as follows:

$$I(h;i) = \log_2 \left(P(h \mid i) / P(h) \right)$$

The logarithm to the base 2 is used so that the units of $I(h;i)$ are bits of information. The converse relation, which shows what information is provided about event i when event h is found to occur, is as follows:

$$I(i;h) = \log_2 \left(P(i|h) / P(i) \right)$$

$I(i;h)$ is identical to $I(h;i)$, which is why this measure of information is called the mutual information between the two events. The joint probability of h and i is $P(h,i)$, which is the probability of both events occurring. If h and i are single events within two sequences of events called H and I, we can calculate mutual information for each event in H compared in turn with each event in I. The average mutual information for the entire sequence pair is found by multiplying the joint probability of every possible event pair within those sequences by the mutual information of that event pair, then finding the grand total of these products. It is also possible to calculate the variance of the mutual information when examining the relation between two entire sequences. Sharman describes a special case of mutual information which occurs when the occurrence of a given outcome i uniquely specifies the outcome of event h. In such a case $P(h|i) = 1$, and

$$I(h;i) = \log_2(1 / P(h)) = - \log_2 P(h) = I(h)$$

This is said to be the **self information** of an event, and is equivalent to the entropy described by Shannon. Another link between mutual information and the information described by Shannon is given by Sneath and Sokal (1973), who provide the following formula:

$$I(h) + I(i) - I(h,i) = I(h;i)$$

Joint information $I(h,i)$ is said to be the union of the information content of two characters in a sequence, while mutual information, $I(h;i)$ is their intersection. The mutual information of two events, such as two words in a text, h and i, is also given by the formula

$$I(h;i) = \log_2 \left(P(h,i) / P(h).P(i) \right)$$

where $P(h,i)$ is the probability of observing h and i together, and $P(h)$ and $P(i)$ are the probabilities of observing h and i anywhere in the text, whether individually or in conjunction. If h and i tend to occur in conjunction, their mutual information will be high. If they are not related and occur together only by chance, their mutual information will be zero. Finally, if the two events tend to 'avoid' each other, such as consecutive elements in a forbidden sequence of phonemes, mutual information will be negative.

The term 'mutual information' is generally used in computational linguistics as described in this section, but strictly speaking the term **specific mutual information** should be used (Smadja, McKeown and Hatzivassiloglou 1996). This is to distinguish it from **average mutual information**, of which specific mutual information constitutes only a part. Specific mutual information is given by the formula

$$I(X,Y) = \sum_{x \in \{0,1\}} \sum_{y \in \{0,1\}} p(X = x, Y = y) \log_2 \frac{p(X = x, Y = y)}{p(X = x)p(Y = y)}$$

Specific mutual information $I(X,Y)$ is simply the logarithm of the probability of the two events X and Y occurring together divided by the probability of the two events whether together or in isolation. In order to calculate average mutual information we must also add on the logarithms of the following quantities:

1. the probability of the first event occurring when the second event does not occur divided by the product of the independent probabilities of the first event occurring and the second event not occurring
2. the probability of the second event occurring when the first event does not occur divided by the product of the independent probabilities of the second event occurring and the first event not occurring
3. the probability of both events simultaneously not occurring divided by the product of the independent probabilities of the first event not occurring and the second event not occurring

Practical applications of specific mutual information include the derivation of monolingual and bilingual terminology banks from corpora, as described in Chapter 4, Sections 3.2.4 to 3.2.6, the segmentation of undelimited streams of Chinese characters into their constituent words as described in Chapter 4, Section 3.2.11, and the identification of idiomatic collocations (see this chapter, Section 4.7).

2.8 Stochastic processes: a series of approximations to natural language

Examples of stochastic processes provided by Shannon are natural written languages, such as Chinese, English or German, continuous information sources that have been **quantised** and rendered discrete, such as quantised speech, and mathematical models where we merely define abstractly a stochastic process which generates a sequence of symbols. The simplest such model is one where the symbols are independent and equiprobable, such as successive trials of a dice or the entries in a random number table. **This is the zero–order approximation**. Using such a model, Shannon obtained the following sample of text:

xfoml rxkhrjffjuj zlpwcfwkcy ffjeyvkcqsghyd

In a slightly more sophisticated model, the symbols are independent but occur with the frequencies of English text. This is called the **first–order approximation**, and might produce the following:

ocro hli rgwr nmielwis eu ll nbnesebya th eei alhenhtppa oobttva nah

In the **second–order approximation**, successive symbols are not chosen independently, but their probabilities depend on preceding letters. This reproduces the bigram structure of English, where the frequencies of adjacent character pairs are based on those found in real text. For this approximation, Shannon obtained the following:

> on ie antsoutinys are t inctore st bes deamy achin d ilonasive tucoowe at teasonare fuzo tizin andy tobe seace ctisbe

In the **third-order approximation**, the trigram structure of English is reproduced:

> in no ist lat whey cratict froure birs grocid pondenome of demonstures of the reptagin is regoactiona of cre.

The resemblance to ordinary English text increases quite noticeably at each of the above steps. Shannon observed that these samples have reasonably good structure (i.e., could be fitted into good sentences) out to about twice the range that is taken into account in their construction. Analogous patterns can also be generated by using words, lemmas or parts-of-speech tags rather than letters as symbols, as in the following first-order word approximation for English (Paducheva 1963, p. 143):

> representing and speedily is an good apt or come can different natural here in came the to of to expert gray come to furnishes the line message had be there

The following is a second-order word approximation for English:

> the head and in frontal attack on an English writer that the character of this point is therefore another method for the letters that the time of who ever told the problem for an unexpected

In the first order case, a choice depends only on the preceding letter and not on any before that. The statistical structure can then be described by a set of **transition probabilities** $P_i(j)$, the probability that the letter i is followed by the letter j. An equivalent way of specifying the structure is to give the bigram or two-character sequence probabilities $P(i,j)$, the relative frequency of the bigram i,j. Pratt (1942) gives the following examples of the frequency of common bigrams per 1000 words of normal text: TH 168, HE 132, AN 92, RE 91 and ER 88. Bigram frequency tables can be normalised into transition probability tables, as shown in the following example. Imagine that the letter T can only be followed by H, E, I, O, A, R and T. The relative frequencies of the resulting bigrams are TH 168, TE 46, TI 45, TO 41, TA 29, TR 29 and TT 9. First find the sum of these relative frequencies, which is 367. The probability of encountering the letter H if the previous letter was T is then 168/367 or 0.46.

The next increase in complexity involves trigram frequencies. For this we need a set of trigram frequencies $P(i,j,k)$ or transition probabilities $P_{ij}(k)$. Pratt gives the following examples of English trigram frequencies: THE 5.3 per cent, ING 1.6 per cent, ENT 1.2 per cent, ION 1.2 per cent, ... , AAA 0 per cent,

AAB 0 per cent. It is interesting to note the extent to which natural languages are characterised by their highest-frequency trigrams. For German the commonest trigrams are *EIN, ICH, DEN, DER*, while for French they are *ENT, QUE, LES, ION*. The most common in Italian are *CHE, ERE, ZIO, DEL*, and in Spanish they are *QUE, EST, ARA, ADO*. As a general case, we talk about an *n*-gram model. For practical purposes, when producing approximations to natural language, a trigram model is probably enough. Above this level, we are faced with the law of diminishing returns and very large, sparse transition matrices. The work we have covered so far is bringing us closer and closer to the relevance of language modelling to corpus linguistics. The ability of the corpus to provide data to language models allows them in turn to process corpora. But before we see exactly how this is done, we need to consider a few more aspects of language modelling.

2.9 Discrete Markov processes

Stochastic processes of the type described in the previous section are known as **discrete Markov processes**. The theory of Markov processes was developed by A. A. Markov (1916), as a result of his study of the first 20,000 words of Pushkin's novel in verse *Eugene Onegin*. Shannon describes a discrete Markov process as consisting of (a) a finite number of possible states, and (b) a set of transition probabilities, $P_i(j)$, which give the probability that if the system is in state S_i it will next go to state S_j. To make this Markov process into an information source, we need only assume that a letter (or part of speech if these are the units of our model) is produced for each transition from one state to another. The states will correspond to what Shannon describes as the **residue of influence** from preceding letters.

In a Markov model, each succeeding state depends only on the present state, so a Markov chain is the first possible generalisation away from a completely independent sequence of trials. A complex Markov process is one where the dependency between states extends further, to a chain preceding the current state. For example, each succeeding state might depend on the two previous states. A Markov source for which the choice of state depends on the *n* preceding states gives an $(n+1)$th-order of approximation to the language from which the transition probabilities were drawn and is referred to as an *n*th-order Markov model. Thus, if each succeeding state depends on the two previous states, we have a second-order Markov model, producting a third-order approximation to the language.

Ergodic Markov processes are described by Shannon as processes in which every sequence produced of sufficient length has the same statistical properties such as letter frequencies and bigram frequencies. In ergodic Markov models every state of the model can be reached from every other state in a finite number of steps. Natural language is an example of an ergodic Markov process.

2.10 Hidden Markov models

Rabiner (1989) describes how the theory of Markov models can be extended to include the concept of hidden states, where the observation is a probabilistic function of the state. In observable Markov models, such as those described by Shannon, each state corresponds to an observable event. However, according to Dunning (CORPORA list[2]), for a **hidden Markov model** (HMM), the output is not the internal state sequence, but is a probabilistic function of this internal sequence. Dunning also states that some people make the output symbols of a Markov model a function of each internal state, while others make the output a function of the transitions. Jakobs (CORPORA list) describes that for the hidden Markov model, instead of emitting the same symbol each time at a given state (which is the case for the observable Markov model), there is now a choice of symbols, each with a certain probability of being selected.

Rabiner explains the operation of a hidden Markov model by reference to the **urn and ball model**. Imagine a room filled with urns, each filled with a mixture of different coloured balls. We can move from one urn to another, randomly selecting a ball from each urn. The urns correspond to the states of the model, and the colours of the balls correspond to the observations, which will not always be the same for any particular urn. The choice of urns to be visited is dictated by the state transition matrix of the hidden Markov model. The model is said to be hidden because *we only see the sequence of balls which were selected*, not the sequence of urns from which they were drawn.

A hidden Markov model is a doubly stochastic process which consists of (a) an underlying stochastic process that cannot be observed, described by the transition probabilities of the system, and (b) a stochastic process which outputs symbols that can be observed, represented by the output probabilities of the system (Sharman 1989). The essential components of a hidden Markov model (where the entire model is denoted by λ) can be summarised by the entire set of transition probabilities (denoted by A), the entire set of output probabilities (denoted by B) and its initial state (denoted by π).

Sharman (1989) states that when hidden Markov models are used in real-world applications, three important problems that must be solved are evaluation, estimation and training. The evaluation problem is to calculate the probability that an observed sequence of symbols occurred as a result of a given model, and may be solved using the **forward–backward algorithm**, described in Section 2.10.1. In the estimation problem, we observe a sequence of symbols produced by a hidden Markov model. The task is to estimate the most likely sequence of states that the model went through to produce that sequence of symbols, and one solution is to use the **Viterbi algorithm**, described in Section 2.10.2. During training, the initial parameters of the model are adjusted to maximise the probability of an observed sequence of symbols. This will enable the model to predict future sequences of symbols. Training may be performed by the **Baum–Welch re-estimation procedure**, described in Section 2.10.3.

2.10.1 Evaluation: the forward–backward algorithm

For an HMM with known parameters, and for the particular sequence of states the model has passed through, the probability of the entire sequence of observed symbols produced by the model is simply the product of the probability of the first observation generated by the first state multiplied by the probability of the second observation arising from the second state, then multiplied by the probability of each observation returned by each state in the sequence in turn, up to the probability of the final observation produced by the final state. This is described by the formula

$$\Pr(O|I,\lambda) = b_{i_1}(O_1) \times b_{i_2}(O_2) \times \dots \times b_{i_T}(O_T)$$

O is the sequence of observed symbols, I is the sequence of states undergone by the model, and λ is the model itself. Thus, $\Pr(O|I,\lambda)$ is the probability of encountering the sequence of observations given the sequence of states of the model and the model itself. While I is used to denote the entire sequence of states that the model has passed through, each individual state is denoted i, where i_1 is the first state, i_2 the second, and so on. $b_{i_1}(O_1)$ is the probability of observing a symbol when the model is in the first state, $b_{i_2}(O_2)$ is the probability of observing a symbol when the model is in the second state, and since T is the number of states in the entire sequence, $b_{i_T}(O_T)$ is the probability of observing the final symbol in the final state. The three dots (...) mean that we must also multiply by the probabilities of observing all symbols for all states between the second and final state. b is used to denote each individual output probability because B was used to denote the full set of output probabilities possessed by the model.

The probability of a given sequence of states (I) that a model (λ) goes through depends only on the initial state and the transition probabilities from state to state. This is shown by the formula

$$\Pr(I|\lambda) = \pi_{i_1} \times a_{i_1 i_2} \times a_{i_2 i_3} \times \dots \times a_{i_{T-1} i_T}$$

Since π denotes the initial state, one of the parameters of the model, π_{i_1} denotes the a priori probability of the first state of the model. π_{i_1} is first multiplied by $a_{i_1 i_2}$ (the probability of making the transition from state one to state two) then multiplied by $a_{i_2 i_3}$ (the transition probability for moving from the second state to the third state) then multiplied in turn by all the other transition probabilities for moving from one state to the next, and finally by the transition probability of moving from the penultimate state to the final state. Thus, the initial probabilities start the model off and the transition probabilities keep it stepping.

For a given model, the joint probability of the sequence of observations O and the sequence of states I occurring together is the product of the two quantities derived above and is represented as:

$$\Pr(O, I|\lambda) = \Pr(O|I,\lambda) \times \Pr(I|\lambda)$$

The overall probability of the observation sequence Pr(O) is the sum of all possible different state sequences which could have given rise to the observed sequence of symbols

$$\Pr(O)=\sum_I \Pr(O|I,\lambda)\times \Pr(I|\lambda)$$

where the summation symbol Σ with subscript I, pronounced 'Sum over all I', means that the product $\Pr(O|I,\lambda) \times \Pr(I|\lambda)$ must be found for all possible values of I (state sequences), then added together. Combining all three formulae derived in this section we obtain

$$\Pr(O)=\sum_{i_1,i_2,\dots i_T} \pi_{i_1}\times b_{i_1}(O_1)\times a_{i_1 i_2}\times b_{i_2}(O_2)\times \dots \times a_{i_{T-1}i_T}\times b_{i_T}(O_T)$$

The summation symbol Σ with subscript i_1, i_2, ..., i_T means that the following product must be found for all possible combinations of states, and the results added together to produce an overall probability value for the observed sequence of symbols. In fact this requires too many calculations to be computationally feasible,[3] so a more efficient method of calculating the observed sequence is required. Such a method is the forward–backward algorithm.

The forward–backward algorithm consists of two distinct phases, a forward pass and a backward pass. In the forward pass, the initial state probabilities of the model are used to calculate subsequent state probabilities, while in the backward pass, the final state of the model is used as the starting point, from which one calculates back to find earlier state probabilities. Either the forward pass or the backward pass can be used to calculate the probability of the observed sequence of symbols produced by a hidden Markov model.

In the forward pass, the value $\alpha_t(i)$ is the probability of the partial sequence of observations up to time t, which results in being in a given state denoted $q(i)$ for the model λ. This is shown formally as:

$$\alpha_t(i)=\Pr(O_1,O_2,\dots,O_t,i_t=q_i|\lambda)$$

For each possible state of the model i, the probability of being in state i at time $t=1$ (the initial state) and producing the first symbol, is first calculated using the formula

$$\alpha_1(i)=\pi_i\times b_i(O_1)$$

This means that the probability of observing the first symbol is the a priori probability of being in the state that produced the first symbol multiplied by the probability that that state would produce the observed symbol. The probability of being in state $q(i)$ at time t and emitting the symbol O_t is then calculated for each successive time step from 1 to T, according to the formula:

$$\alpha_t(i)=\sum_{j=1}^{N}\left[\alpha_{t-1}(j)\times a_{ji}\right]\times b_i(O_t)$$

$\alpha_{t-1}(j)$ is the probability of the partial sequence of observations up to time $t-1$, which results in the model being in state $q(j)$. The $j = 1$ below the summation sign Σ and the N above it shows that $\alpha_{t-1}(j)$ must be calculated for all possible states j between 1 and N, where N is the number of possible states. The summation must be performed over all preceding states, since in theory any state could cause a transition to the current state. Each time $\alpha_{t-1}(j)$ is calculated, it must be multiplied by α_{ji} which is the probability of the transition from state j to state i. These products are all added together, then multiplied by $b_i(O_t)$, the probability that state i would produce the observed symbol at time t. This type of equation is called **recursive**, since the state probability at one time is defined in terms of the state probabilities at the previous time step. Having found the α values for time t, we can substitute them in the above formula for the α values at time $t-1$, and in this way calculate the α values for the next time step, $t+1$. This process is repeated until we obtain $\alpha_T(i)$, the probability of state i being the final state. To commence the process, use the formula for $\alpha_1(i)$, then use this as the first value of $\alpha_{t-1}(j)$ in the formula for $\alpha_t(i)$.

Finally, to calculate the overall probability of all the observations given the model, from all the possible final states, we must find all values of $\alpha_T(i)$ for all values of i in the range 1 to N, then add them together, as shown in the equation below:

$$\Pr(O|\lambda) = \sum_{i=1}^{N} \alpha_T(i)$$

The backward pass is calculated in analogous fashion to the forward pass. For the backward pass, we define $\beta_T(i)$ as the probability of the partial sequence of observations starting at time t, which results in the model being in the state $q(t)$ for a given model λ:

$$\beta_T(i) = \Pr\left(O_{t+1}, O_{t+2}, \ldots, O_T, i_t = q_i | \lambda\right)$$

For each state i in the range 1 to N the starting value of being in state i at time $t = T$ is calculated, using the relation $\beta_T(i) = 1$. This relation holds because there must be some path by which the model gets to the final state. For the backward pass, a recursive formula is used to calculate the probability of being in state $q(i)$ at time and producing the observed symbol O_t for each previous time step, $T-1$, then $T-2$, and so on back to $t = 1$. The formula is

$$\beta_t(i) = \sum_{j=1}^{N} \left[\beta_{t+1}(j) \times a_{ij}\right] \times b_i(O_t)$$

Finally, to calculate the overall probability of all the observations given the model, from all the possible final states, we must find all values of $\beta_1(i)$ for all values of i in the range 1 to N, then add them together, as shown in the following equation:

$$\Pr(O|\lambda) = \sum_{i=1}^{N} \beta_1(i)$$

2.10.2 Estimation: the Viterbi algorithm

In the estimation problem, we wish to find the most likely sequence of states for a sequence of observations produced by a given model. For example, it is possible to create a Markov model describing all possible parses of a sentence where some of the constituent words could have more than one part of speech assigned to them. The single most likely parse can be found by means of the Viterbi algorithm. The **Viterbi algorithm** conceptually resembles the forward pass algorithm described in the previous section but, instead of summing the transition probabilities which lead to a given state, the top M probabilities are taken instead. Since these top M most probable state sequences are used as the basis for calculating the transition probabilities which lead to the next state, the Viterbi algorithm is also a form of dynamic programming, which is described in Chapter 3, Section 4.6.

In its most general form, the Viterbi algorithm may be viewed as a solution to the problem of estimating the state sequence of a discrete-time finite-state Markov process. The MAP (maximum a posteriori probability) estimation problem is formally identical to the problem of finding the shortest route through a certain graph. Forney (1973) describes the process graphically by considering a trellis, where each node corresponds to a distinct state at a given time (which in the case of a part-of-speech tagger would be a candidate tag for a given input word) and each branch represents a transition to some new state through the state diagram. The trellis begins and ends at known states, which in the case of a part-of-speech tagger would be the null tags at the start and end of each sentence. To every possible state sequence there exists a unique path through the trellis and vice versa. Figure 2.2 shows a Markov model for part-of-speech tagging viewed as a trellis.

Every path in Figure 2.2 (such as those leading from the initial null tag to the final null tag) may be assigned a length proportional to $-\log_e$ (probability of that path being correct). To find the state sequence for which the probability is maximum, the algorithm finds the path whose length is minimum. The length

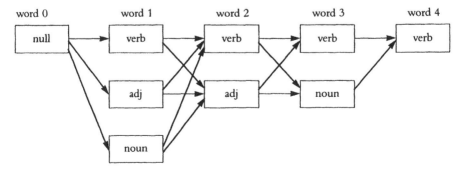

Figure 2.2 Markov model for part-of-speech tagging viewed as a trellis

of a sequence of transitions is the sum of the individual lengths of those transitions. In a part-of-speech tagger, the path length of an individual transition might be given by a function such as $-\log_e((p(i) \times p_i(j, k)) + c)$; where $p(i)$ is the a priori probability of a term being assigned a given part-of-speech tag $p_i(j, k)$, is the probability of the trigram k, j, i occurring in sequence and c is a small constant to prevent path lengths of infinity occurring.

For any particular time, from one state to another there will generally be several path segments, the shortest M of which are called **survivors**. Thus, at any one time k, one need remember only the M survivors and their lengths. To get to time $k+1$, one need only extend all time k survivors by one time unit, compute the lengths of the extended path segments, and for each node $x(k+1)$ select the shortest extended path segment terminating in $x(k+1)$ as the corresponding time $(k+1)$ survivor. This recursive process continues indefinitely without the number of survivors ever exceeding M. Although the extension to infinite sequences is theoretically possible, any practical implementation must select a maximum sequence length, and a value of M must be specified.

2.10.3 Training: the Baum–Welch algorithm

No analytical solution is known for the problem of training a hidden Markov model. Thus we must employ iterative techniques such as the Baum–Welch re-estimation algorithm. The task is to adjust the model parameters to maximise the probability of an observed sequence of symbols (Sharman 1989).

Given a model which produces an observed sequence of symbols, we wish to find $\xi_t(i, j)$, the probability of being in state q_i at time t and making a transition to state q_j at time $t+1$. Then:

$$\xi_t(i, j) = \Pr\left(i_t = q_i, i_{t+1} = q_j \mid O, \lambda\right) = \frac{\alpha_t(i) \times a_{ij} \times b_j(O_{t+1}) \times \beta_{t+1}(i)}{\Pr(O\lambda)}$$

The symbols used in this equation are as follows: $\alpha_t(i)$ is the probability of arriving in state q_i at time t by any path leading from the initial state, and producing the output symbol O_t. This is calculated for the forward pass of the forward–backward algorithm. a_{ij} is the probability of making the transition from state q_i to q_j. The transition probabilities are original parameters of the model. $b_j(O_{t+1})$ is the probability of producing the output symbol at the next time step, O_{t+1}. The output probabilities are also original parameters of the model. $\beta_{t+1}(i)$ is the probability of leaving state q_j at time $t+1$ by any path, and eventually getting to the final state. This is calculated by the backward pass of the forward–backward algorithm.

The probability of being in state q_i at time t is called $\gamma_t(i)$ and is found by adding together all the values of $\xi_t(i, j)$ calculated for all values of i from 1 to N, the total number of states in the model, as follows:

$$\gamma_t(i) = \sum_{i=1}^{N} \xi_t(i, j)$$

The expected number of transitions made out of state q_i is called Γ_i, which is the sum of all values of $\gamma_t(i)$ calculated at every time step from $t = 1$ to $t = T$, where T is the total number of steps taken by the model, as shown by the equation

$$\Gamma_i = \sum_{t=1}^{T-1} \gamma_t(i)$$

The expected number of transitions made from state q_i to state q_j is called Ξ_{ij}, and is the sum of all values of $\xi_t(i,j)$ taken at each time step from $t = 1$ to $t = T$:

$$\Xi_{ij} = \sum_{t=1}^{T-1} \xi_t(i,j)$$

In order to optimise the parameters of the model to maximise the probability of the observation sequence, we must re-estimate the values of the three parameters defining the model; namely, the initial state probabilities, the transition probabilities and the output probabilities. First we re-estimate the probability of each of the initial states. The original probability of the model initially being in state i was called π_i, and the re-estimated probability is called $\bar{\pi}_i$. The $\bar{\pi}_i$ values are equal to the $\gamma_1(i)$ values, which are the values of $\gamma_1(i)$ when $t = 1$. Secondly, the new estimate of each state transition probability, called $\bar{a}_{ij}$, is found using the relation

$$\bar{a}_{ij} = \frac{\Xi_{ij}}{\Gamma_i}$$

This is the ratio of the expected number of transitions from one state to the next, divided by the total number of transitions out of that state. Finally, the new estimate of each output probability, called $\bar{b}_i(k)$, is the ratio of the expected number of times of being in a state and observing a given symbol divided by the expected number of times of being in that state, given by the formula

$$\bar{b}_i(k) = \frac{\Xi_{ij}(k)}{\Gamma_i}$$

We now have a new model, $\bar{\lambda}$, which is defined by the re-estimated para-meters $\bar{A}, \bar{B},$ and $\bar{\pi} i$. Thus

$$\bar{\lambda} = (\bar{A}, \bar{B}, \bar{\pi})$$

These values may be used as the starting points of a new re-estimation procedure, to obtain parameter estimations which account even better for the observed sequence of symbols. By continuing this iterative process, we will eventually reach a point where the re-estimated parameters are no longer any different to the input parameters. The values are then said to have converged. The convergence point is called a local maximum, which does not preclude the possibility that the algorithm may have missed an even better set of parameters called the global maximum. If the local maximum is the same as the global maximum, we have found the set of model parameters which most closely accounts for the sequence of observed symbols. The Baum–Welch algorithm is an example of a class of algorithms called estimation–maximisation algorithms or EM-algorithms, all of which converge on a local maximum. Highly mathematical accounts of the use of EM-algorithms are given by Baum and Eagon (1967), Baum et al. (1970) and Dempster, Laird and Rubin (1977).

2.11 Perplexity

Perplexity is the final concept from information theory that we will consider. Sharman (1989) writes that when considering the perplexity of a text, we are asking: what is the size of the set of equiprobable events which has the same information? For example, perplexity is the average size of the set of words between which a speech recogniser must decide when transcribing a word of the spoken text. The difficulty of the task of guessing the unknown event (the next word) is as difficult as guessing the outcome of a throw of a dice with 2^H faces, where H is the information or entropy of that juncture of the word sequence. It is a measure of the average branching of the text produced by a model, and is maximum if the words of the text are randomly selected with uniform probability.

Jelinek (1985) advocates the use of perplexity as an objective measure of language model quality. Proposed changes to a model can then be evaluated by examining whether they succeed in reducing the perplexity of the model. The symbols $w_1, w_2, \ldots, w_n$ denote the sequence of words in a corpus to be recognised, $P(w_1, w_2, \ldots, w_n)$ is the actual probability of that sequence and $\hat{P}(w_1, w_2, \ldots, w_n)$ is the estimate of the probability of that word sequence according to a language model. The amount of information per word in the corpus is estimated by

$$H = -(1/n) \log_2 P(w_1, w_2, \ldots, w_n)$$

where n is the size of the corpus. Since the actual probabilities of the strings of the natural language cannot be known in practice, we must use the estimations of these real probabilities given by a language model. The information per word of language produced by the model is measured by its log probability, which is given by the analogous formula

$$LP = -(1/n) \log_2 \hat{P}(w_1, w_2, \ldots, w_n)$$

The perplexity of the model, PP, is equal to 2^{LP} or

$$[\hat{P}(w_1, w_2, \ldots, w_n)]^{(-1/n)}$$

For natural language, $PP = 2^H$.

2.12 Use of perplexity in language modelling

Sekine (1994) found that measuring the perplexity of texts was a good method of identifying those texts which were written in a genre-specific sublanguage. Such texts tend to use a smaller vocabulary than texts selected randomly from across an entire corpus, and thus tend to have lower perplexity. Sekine's work is described in more detail in Chapter 3, Section 6.

Jelinek (1985) describes the use of the forward–backward algorithm with a hidden Markov chain model for speech recognition to perform the following tasks:

1. to determine the weights used in a formula for the estimation of trigram probability
2. to annotate text automatically by parts of speech
3. to estimate the probabilities in a word prediction formula that involves the part-of-speech classification of words
4. to automatically infer any new part-of-speech classes.

The approach listed first was to overcome the problem that many trigrams which do not occur at all in the training text, do sometimes appear in test sets. The trigram frequencies were smoothed according to a formula which also takes into account the bigram frequency of the second and third word and the a priori probability of the third word. The formula listed third stated that the probability of encountering word 3 after word 1 and word 2 is the probability k of word 3 being the word given its part-of-speech category multiplied by the probability h of part-of-speech tag $g3$ following tags $g1$ and $g2$. Jelinek used perplexity as a measure to show whether the refinement suggested in 1. and 3. improved the original simple trigram model.

To automatically infer part-of-speech categories without making use of any knowledge of existing part-of-speech categories, the M most frequent words of the vocabulary were each assigned to different part-of-speech classes called nuclear classes. Each of the remaining words which do not form class nuclei are originally allowed to belong to any of the M classes. Through a self-organised clustering process the words progressively abandon most of their class memberships until no word belongs to more than a few classes.

3 PROBABILISTIC VERSUS RULE-BASED MODELS OF LANGUAGE

Sampson (1987) states that the probabilistic approach to language analysis is characterised by the use of analytic techniques which depend on statistical properties of language structure rather than reliance on absolute logical rules. The use of statistics enables one to study authentic data drawn from unrestricted domains of discourse rather than preselected domain-specific texts.

For the task of text parsing, an alternative approach to the use of statistical techniques is to use a form of generative grammar, a system of rules which define all possible grammatically correct sentences of the language to be parsed. This produces an 'all or none' situation, where any sentence not conforming to the constraints of the grammar will be deemed incorrect. Examples of generative grammars are generalised phrase structure grammars and augmented transition networks, which are both described by Salton and McGill (1983). However, Sampson points out that no existing generative grammar can account for all the diversity and peculiarities found in authentic natural language. For a fuller discussion of his views, see McEnery and Wilson (1996).

Paducheva (1963) also describes how language studies constantly show instances where phenomena cannot be described fully yet briefly, since language is a system composed of a large number of diverse objects interacting according to very complex laws. It is only rarely that one can formulate strict, fully determined rules about language objects. The application of fully determined rules means that upon realisation of a fully determined complex of conditions, a definite event must take place. For example, it is difficult to formulate strict rules about the use of the article before the noun in English. An analysis of the rules given in ordinary grammar books will show that they do not allow for the unique definition of all the conditions for choosing an article. If the number of factors taken into account is increased, the number of errors will decrease, but the set of rules will become ever more cumbersome, and potentially inaccurate. This suggests that rather than try to define a comprehensive set of fully determined rules, it may be better to say that 'given the occurrence of a set of conditions, an event will occur on a certain proportion of occasions'. Paducheva writes that, because of the complexity, multiplicity and close overlapping of laws governing a natural language, the laws of language codes cannot be described in the same way as the rules for technical codes. Unlike a technical code, the laws of a natural language code can be broken by, for example, the utterances made by people not familiar with the domain, foreign words (which can break the rules of phoneme combination), and new words.

Sapir (1921) said 'All grammars leak', meaning that certain acceptable linguistic combinations will always be overlooked, since the production of an entirely watertight grammar would require the incorporation of an infinite number of rules. The probabilistic approach, on the other hand, allows us to dispense with a grammar of 'all or none' rules. No absolute distinctions between grammatical and ungrammatical forms are made, but instead the likelihood of any linguistic form occurring is described in terms of its relative frequency, thus enabling more than a two-step ranking of sentences. Paducheva concludes that 'Naturally it is better to have such statistical rules ... true no less than 94 per cent of the time, than to have no rules at all or to have 10 pages of rules for each individual object of linguistic study', while Halliday (1991) reminds us that a human parser does not attain 100 per cent accuracy either. This is partly because some instances are inherently indeterminate but also because humans make mistakes. Outside corpus linguistics, such as in physics, it is also convenient to use a probabilisitic model rather than trace a long, involved derivation of a system of equations, as noted by Damerau (1971).

Halliday points out that it has long been accepted that different words have different relative frequencies, as predicted by Zipf's law described in the previous section. For example, native speakers of English would know that the word *go* is more frequent than *walk* which in turn is more frequent than *stroll*. This not only shows that humans are able to internalise vast amounts of probabilistic knowledge pertaining to language, but also leads onto the suggestion that if grammar and lexis are complementary perspectives forming a continuum, models of language should accept relative frequency in grammar just as they should accept relative frequency in lexis. It has been shown that the relative frequencies of grammatical choices can vary both in diachronic studies and across registers, as in Svartvik's (1966) study of the ratio of active to passive clauses in various text genres. Thus, for studies of register variation, probability is of central importance.

Just as native speakers are inherently aware of the relative frequencies of *go*, *walk* and *stroll,* they can do the same for many aspects of grammar, recognising, for example, that the active voice is more common than the passive voice. In other instances they will be less certain, such as in comparing the relative frequencies of *this* and *that*, suggesting that this pair tend to be more equiprobable. Frequency information from a corpus can be used to estimate the relative probabilities within any grammatical system, as shown by Leech, Francis and Xu (1994) to demonstrate the gradience of preference for alternative forms of the genitive rather than the existence of hard and fast rules to state when one form should always be used. Pure all-or-none rule-based systems do not take into account the relative frequency of the allowable linguistic constructs.

Even at the level of pragmatics, probability, rather than hard and fast rules, seems to be used by humans. McEnery (1995) carried out experiments that showed inferencing in language is a matter of degrees of belief rather than absolute belief. His experiments show that language comprehension is probabilistic and not deterministic.

Language may be described on many levels – the level of phonemes, syllables, morphemes, words or letters, and so on. The combination of units on one level is affected by limitations reflecting laws of combinability not only on that level, but on other levels as well. If we wish to describe the laws governing com-bination in the form of structural, qualitative, determinate rules, we must separate the limitations affecting the laws of one level from all the others. In practice it is very difficult to define exactly the criteria of separation. Statistical laws of combination, on the other hand, can be made explicit from immediate observation of corpus texts. One does not have to develop criteria for idealised text as required to divide limitations among levels.

A number of experiments have been performed to test one or more consequences of the Markov model. Looking at some of these studies, we may conclude that some aspects of human speech behaviour appear to be rule directed, while other aspects appear to be probabilistic. Damerau (1971) describes how one class of experiments categorised under hesitation studies showed that observed human behaviour was in accordance with that predicted by the Markov model, while the results of other experiments could be interpreted in the light of either a probabilistic or a rule-based model.

The hesitation studies examine Lounsbury's (1965) hypotheses that

1. hesitation pauses in speech correspond to the points of highest statistical uncertainty in a sequence of words
2. hesitation pauses frequently do not fall at the points where a rule-based grammar would establish boundaries between higher-order linguistic units.

All the hesitation studies used a corpus of recorded speech material. The criteria for identifying hesitation points were either the agreement of two judges (the approach of Tannenbaum, Williams amd Hillier 1965) or the presence of a quarter-second silence period (used by Goldman-Eisler 1958). The **transition probabilities** (the probabilities of two units of language such as words occurring one after the other) which reflect the degree of statistical uncertainty at each point in the corresponding text were estimated using human subjects, who had to guess which words were missing from the text. The transition probabilities were the ratio of correct guesses to total guesses. Tannenbaum, Williams and Hillier used the **cloze procedure** with every fifth word deleted, while Goldman-Eisler told the subjects to guess the next word when shown all previous words, as in Shannon's game, described in Section 4.4. In their examination of Lounsbury's second hypothesis, Maclay and Osgood (1959) showed that approximately half of all pauses occurred within phrases rather than at phrase boundaries, and thus the predictions of the Markov model appeared to be confirmed.

Other studies described by Damerau include that of Miller (1950) who showed that certain statistical approximations to real English (called fourth- and fifth-order Markov models) could be remembered as easily as a meaningful text passage, and that of Somers (1961) who calculated the amount of information in Bible texts. In these experiments, results obtained using one theory can generally be utilised in the context of another theory.

The use of the probabilistic approach may be illustrated by a description of the CLAWS word-tagging system which is based on a Markov model. The function of one version of CLAWS is to assign one of 133 possible parts-of-speech or grammatical tags to each word in the sequence of English text that it is presented with (Garside 1987). A list of candidate tags and their probabilities is assigned to each word using a lexicon. One of these candidate tags is chosen as the most likely candidate for each word, by taking into account both its a priori probability found in the lexicon and a stored matrix of the frequencies with which one word tag tends to follow another in running text. These values are called transition frequencies, and are found empirically from a part-of-speech tagged corpus beforehand. Finally, the most likely sequence of tags in the input sentence is found using the Viterbi algorithm. The CLAWS system will be described in detail in Section 4.1.1.

The CLAWS probabilistic tagging system runs at a rate of between 96 per cent and 97 per cent of authentic text correctly tagged. This result is achieved without using any grammatical knowledge (in the generative grammarian's sense) at all. All its 'knowledge' is probabilistic, being stored in the 133x133 matrix of tag transition frequencies. No cell of this matrix is given a value of 0, so for CLAWS nothing is ruled out as impossible. The CLAWS approach did not set out to achieve psychological plausibility, but, ironically, it is more psychologically plausible than a generative grammar-based system which returns no output at all in response to ill-formed input. Consider what might first occur to you if you see a sentence which contains a word with an unexpected part-of-speech tag, such as 'He waits for a *describe*'. You would notice not only that the word does not usually have that tag but that in view of its context, one might expect it to have that tag. CLAWS would do exactly the same, while a more deterministic system may reject this as ill-formed input.

There are still a number of theoretical problems with the use of probabilistic models of language. The simple Markov process gives a poor approximation of real text, since cohesive features in language such as anaphora may act at great distances. An extreme example of this is Proust's lengthy *A la recherche du temps perdu*, where the very first word was deliberately chosen to be identical to the very last word. The statistical approach to describing combinatory capabilities also possesses the restriction that it does not reflect the qualitative diversity of relations among elements, such as the multiplicity of grammatical relations among words. Similarly, Paducheva points out that language communications can be studied on several levels, corresponding to phonemes, syllables, words and so on. The combination of units on one level is affected by limitations not only on that level, but on other levels as well, and it can be difficult to separate the limitations which apply at one level from those affecting other levels.

4 USES OF INFORMATION THEORY IN NATURAL LANGUAGE PROCESSING: CASE STUDIES

4.1 Part-of-speech taggers

Our principal case study for this chapter is the CLAWS word-tagging system, described by Garside (1987), which performs probabilistic part-of-speech tagging. CLAWS uses the Viterbi algorithm to infer the most likely sequence of actual word tags given all sequences of possible word tags. This task is also performed by the Cutting tagger, which will be described in Section 4.1.2.

While the CLAWS tagger derives its transition probabilities from the Brown corpus, the Cutting tagger can be trained on any text using the Baum–Welch algorithm. Following our examination of the Cutting tagger, we will look at the use of information theory in a variety of relevant fields.

4.1.1 The CLAWS word-tagging system: a corpus annotation tool

Any automatic parsing approach must take into account both that there are a large number of homographs, or words spelt alike but with different meanings, and that natural language is open-ended, developing new words or new meanings at any time. The CLAWS tagging system consists of five separate stages applied successively to a text to be tagged. The first step is pre-editing where the text is cleaned and verticalised (one word is printed above another). This is followed by candidate tag assignment, where each possible tag that might apply to a word is assigned in descending order of likelihood. Thirdly, multi-word units such as idioms are tagged as single items. The fourth step is tag disambiguation: this stage inspects all cases where a word has been assigned more than one tag, and attempts to choose a preferred tag by considering the context in which the word appears, and assessing the probability of any particular sequence of tags. The final phase is manual post-editing, in which erroneous tagging decisions made by the computer are corrected by human editors.

Tag assignment is performed by a program called WORDTAG. Most words can be assigned a set of possible tags by matching against a stored lexicon. The lexicon contains all function words (in, my, was, that, etc.), and the most frequent words in the open classes noun, verb and adjective. This lexicon accounts for a large proportion (65–70 per cent) of the tagging decisions made at this stage. If the word is not found in the lexicon, its suffix is matched against the suffix list, which contains real suffixes such as -ness (suggesting a noun) and also any word endings which are almost always associated with a certain word class; for example, -mp (suggesting a noun or verb: exceptions such as damp are in the lexicon). The suffix list is searched for the longest matching word ending. If all the previous rules fail, tags are assigned the open classes verb, noun or adjective by default. Very few words are assigned this default tagging and tend to be deviant spellings.

The tag disambiguation program is called CHAINPROBS. It is at this stage that the work we have looked at so far becomes relevant to corpus processing. After the initial tag assignment program has run, every syntactic unit has one or more tags associated with it, and about 35 per cent are ambiguously tagged with two or more tags. Such words must be disambiguated by considering their context, and then re-ordering the list of tags associated with each word in decreasing order of preference, so that the preferred tag appears first. With each tag is associated a figure representing the likelihood of this figure being the correct one, and if this figure is high enough the remaining tags are simply eliminated. Thus, some ambiguities will be removed, while others are left for the manual post–editor to check. In most cases the first tag will be the correct one.

The disambiguation mechanism requires a source of information as to the strength of the links between pairs of tags: much of this information was derived from a sample taken from the already-tagged and corrected Brown corpus. Marshall (1987) reports that a statistical analysis of co–occurring tags in the Brown corpus yielded a transition matrix showing the probability of any one tag following any other, based on the frequency with which any two tags are found adjacent to each other. The values originally used by CLAWS were derived as the ratio of the frequency of the tag sequence 'A followed by B' in the Brown corpus divided by the frequency of the tag A in the Brown corpus, as shown in the following formula:

$$\frac{\text{Frequency of the tag sequence 'A followed by B' in the Brown corpus}}{\text{Frequency of tag A in the Brown corpus}}$$

This transition matrix stored the conditional probabilities of 'tag B given tag A' derived in this way for each pair of tags A, B in the tag set.

Given a sequence of ambiguously tagged words, the program uses these one-step probabilities to generate a probability for each possible sequence of ambiguous tags. Consider a sequence where the second and third words are ambiguously tagged, such as w1 (A), w2 (B or C), w3 (D or E), w4 (F). To disambiguate w2 and w3, we must find the probabilities of the following sequences: A B D F, A C D F, A B E F and A C E F. Marshall gives an example in the sentence *Henry likes stews*. This would receive the following candidate tags from WORDTAG: *Henry* (NP), *likes* (NNS or VBZ), *stews* (NNS or VBZ), . (.). The full stop is given its own unambiguous category. For each of the four possible tag sequences spanning the ambiguity, a value is generated by calculating the product of the frequencies per thousand for successive tag transitions taken from the transition matrix, as shown below:

value(NP-NNS-NNS-.) = 17 x 5 x 135 = 11,475
value(NP-NNS-VBZ-.) = 17 x 1 x 37 =   629
value(NP-VBZ-NNS-.) = 7 x 28 x 135 = 26,460
value(NP-VBZ-VBZ-.) = 7 x 0 x 37 =   0

The probability of a sequence of tags is then determined by dividing the value obtained for the sequence by the sum of the values for all possible sequences. For example, the probability of the likeliest sequence NP-VBZ-NNS-. is 26,460 divided by (11,475 + 629 + 26,460 + 0) or 69 per cent.

To find the probability that an individual word has a given tag, we find the ratio of the sum of the probabilities of all sequences where this word has this tag divided by the sum of the probabilities of all possible tag sequences. Using the *Henry likes stews*. example, the probability of *likes* being tagged as a plural noun is (11,475 + 679) divided by (11,475 + 629 + 26,460 + 0) or 31 per cent.

If the length of the ambiguous word sequence is longer, or the number of alternative parts of speech at each point is greater, it may not be feasible to work out the product of the transition frequencies for every single possible sequence. To cater for such circumstances, the Viterbi algorithm is employed. At each stage in the Viterbi algorithm, only the *n* best paths encountered so far are retained for further consideration. Let us see what happens if we evaluate the *Henry likes stews.* example using the Viterbi algorithm where *n* = 2. The sequence is unambiguous as far as *Henry*, which can only be a proper noun, so this is our starting point. The word *likes* has just two interpretations, so both may be kept as a basis for future calculation. The word *stews* also has two interpretations, meaning that we now have 2 x 2 = 4 possible tag sequences. The likelihood of the sequence NP-NNS-NNS = 17 x 5 = 85, the sequence NP-NNS-VBZ has likelihood 17 x 1 = 17, NP-VBZ-NNS has likelihood 7 x 28 = 196, and NP-VBZ-VBZ has likelihood 7 x 0 = 0. The two sequences with greatest likelihood are therefore NP-NNS-NNS and NP-VBZ-NNS. The next word in the sequence is the unambiguous full stop. Only the best two sequences at the previous stage are now considered, so we calculate the likelihoods of the sequences NP-NNS-NNS-. and NP-VBZ-NNS-. There is no need for us to calculate the likelihoods of the other two possible sequences. Now that we have come to the end of the ambiguous word sequence, we find the likelier of the two most probable tag sequences. Thus, it may be seen that the Viterbi algorithm has the advantage of avoiding a number of calculations which increases with ambiguous sequence length and number of possible tags for each word, but it also has the disadvantage of sometimes missing the overall best sequence or 'global maximum'.

Both modes of calculation can be refined by the use of rarity markers for the candidate tags. A tag occurring less than 10 per cent of the time, labelled @, caused the sequence probability to be divided by two, and a tag occurring less than one per cent of the time, labelled %, caused the sequence probability to be divided by eight. These values were found by trial and error. Although the final version of CHAINPROBS contained further refinements, even the simple algorithm described here, which uses only information about frequencies of transitions between immediately adjacent tags, gave 94 per cent correct tagging overall and 80 per cent correct tagging for the words which were ambiguously tagged. Errors may arise due to

1. erroneous lexicon entries
2. the occasional reliance by WORDTAG on general suffix rules rather than individual words
3. errors from the transition matrix
4. relations between non-adjacent words.

Modifications were made after experimenting with different formulae for the calculation of the transition-matrix values. The final version of CHAINPROBS uses the formula (frequency of the tag-sequence 'A followed by B') divided by (frequency of tag A x frequency of tag B), as shown in the following formula:

Frequency of the tag-sequence 'A followed by B' in the Brown corpus

---

Frequency of tag A in the Brown corpus x frequency of tag B in the Brown corpus

Including the frequency of the second tag in the denominator reduces the tendency of the system to prefer individual high-frequency tags, so that low-frequency tags have more chance of being chosen. A second modification is that no value in the matrix should be zero, so a small positive value is associated with any transition that fails to occur in the sample. This ensures that even for texts containing obscure syntax or ungrammatical material, the program will always perform some analysis. The problem of errors involving relations between non-adjacent words, is exemplified by the fact that the word *not* is less useful in determining the category of the following word than is the word preceding *not*. *not* itself is unambiguously tagged XNOT, so a mechanism has been built into CHAINPROBS which causes *not* to be ignored, so that the words on either side are treated as if they were adjacent. At one stage the model, which is based on the probability of tag pairs, was bolstered by the addition of scaling factors for preferred and dispreferred tag triples. However, the incorporation of such information was found to degrade performance.

A part-of-speech tagger, such as the Cutting tagger or CLAWS, uses context to assign parts-of-speech to words, based on the probabilities of part-of-speech sequences occurring in text incorporated into a Markov model. The CLAWS system is an example of a first-order Markov model applied to the task of word tagging. CLAWS works in both directions, so according to Atwell (1987), when considering the likelihood of a candidate part-of-speech tag for a word, both the immediately preceding tag and the immediately following tag are taken into account. Higher-order models would take even more context into account, but would be computationally expensive.

4.1.2 The Cutting tagger

The implementation of a part-of-speech tagger based on a hidden Markov model which makes use of the Baum–Welch algorithm is described by Cutting et al. (1992). The Cutting tagger models word order dependency. The transition probabilities in its underlying Markov model correspond to the probabilities of part-of-speech tags occurring in sequence. In a probabilistic parsing system such as CLAWS or the Cutting tagger, the observations are the words of the original text, and the set of output probabilities is the probability of the word given its part-of-speech tag. The model is said to be hidden since we are able only to observe directly the ambiguity classes that are possible for individual words, where the ambiguity class of each word is the set of parts of speech that word can take, only one of which is correct in a particular context. The sequence of correct parts of speech given the context cannot be observed directly, so the most probable sequence must be inferred mathematically. The

Cutting tagger can be trained from a relatively small training set of about 3000 sentences, using the Baum–Welch algorithm to produce maximum likelihood estimates of the model parameters A, B and π. The most likely sequence of underlying state transitions (parts of speech) given new observations (the words in a test corpus) is then estimated using the Viterbi algorithm. The advantage of looking for sequences of tags rather than sequences of individual terms is that the number of tags will be much smaller than the number of individual terms, and thus a grammar based on tag sequences will be much simpler than one which considers all terms individually. Some tag sets, including the one used by the Cutting tagger, have additional categories to allow each of the commonest function words to belong to their own class.

4.1.3 Other applications of the Markov model in natural language processing

Many authors have employed Markov models in text recognition. Forney (1973) reports how the contextual information inherent in natural language can assist in resolving ambiguities when optically reading text characters. Hanlon and Boyle (1992) describe how Markov models have been used for the recognition of cursive script. Candidate words with matching 'envelope' features or silhouettes are assigned from a lexicon in a pre-processing stage, and contextual information provided by a Markov model is then used to decide between candidate words. Lyon and Frank (1992), in their work on speech interfaces, make use of a Markov model to identify commonly recurring grammatical sequences.

4.2 Use of redundancy in corpus linguistics

While studying the relative frequencies of complementary grammatical features in a corpus of novels and scientific texts, Halliday (1991) found an interesting bimodal pattern. Pairs of grammatical features seemed either to be close to equiprobable, resulting in redundancy of less than 10 per cent, as in the case of the use of *this* and *that*, or had skewed relative frequencies, occurring in a ratio of about 0.9 to 0.1. For example, the ratio of the use of the active voice to the passive voice was 0.88 to 0.12. This ratio corresponds with redundancy of about 50 per cent.

4.3 An experimental method for determining the amount of information and redundancy: Shannon's game

Shannon (1949) states that everyone who speaks a particular language possesses knowledge of the statistical structure of that language, albeit unconsciously. For example, if someone is asked to guess the next letter in a text when all the preceding letters are known, the guesses will not be entirely random, but will be based on an intuitive knowledge of the probabilities and conditional probabilities of letters.

The approximate value of redundancy in a natural language can be found by the following simple experiment described by Paducheva (1963), which is one version of Shannon's game. A subject is asked to guess each letter in turn, starting with the first letter, in a text of definite length such as 100 letters. If any letter is not guessed correctly, the subject is told the right letter, then goes on to guess the next letter. The space between words is counted as one letter. The proportion of letters guessed correctly enables one to estimate entropy. For example, 50 letters guessed correctly out of a total of 100 letters shows redundancy at around 50 per cent. The same experiment can be performed using words or phonemes.

In another version of Shannon's game, described by Kahn (1966), the subject is also asked to guess the identity of each letter of a text starting with the first letter. This time, however, if the subject guesses incorrectly at any stage, he or she is not told the correct letter, but must keep on guessing until a correct guess is made. In this way we can record how many guesses a subject needed to determine the correct letter in an unknown text. The number beneath each letter in the example of Table 2.1 below gives the number of guesses one subject made.

T H E R E – I S – N O – R E V E R S E – O N – A – M OT OR C Y CL E
1 1 1 5 1 1 21 1 2 1 1 15 1 17 1 1 1 2 1 3 2 1 2 2 7 1 1 1 1 4 1 1 1 1

Table 2.1 Number of guesses made at each point in a text during Shannon's Game

Since the number of guesses required at each juncture reflects the degree of uncertainty or entropy at that point, Shannon's game shows that entropy is lowest at the start of new words and between morphemes.

4.4 Information theory and word segmentation

Juola, Hall and Boggs (1994) describe the MORPHEUS system which segments full words into their constituent morphemes, employing only the statistical properties of information theory. As a morpheme is the smallest unit of language that carries meaning, it may be semantic or syntactic, as in the word *un-happi-ness* which consists of *not* + *happy* + noun suffix. The division of words into their constituent morphemes or morphological analysis assists in corpus annotation (part-of-speech tagging and semantic code assignment) and dictionary construction. Another function is the stripping of derivational morphemes from the ends of words, so all grammatical forms such as singular and plural, present and past tense, of a word can be processed as one for document retrieval. Morpheme identification is difficult. There may be variant spellings of the same morpheme (such as *happy* and *happi-*) called allomorphs, or connecting vowels as in *toxic-o-logy*. One non-statistical method of morpheme identification is dictionary look-up, as described by Oakes and Taylor (1994), but this can be difficult for unfamiliar languages. The importance of identifying derivational morphemes in unknown languages is shown by the case of Linear B, described in Chapter 5, Section 4.1, where the identification of commonly occurring terminal sequences was a key step in the decipherment.

Since entropy is a measure of unpredictability, it is highest at morpheme boundaries. By gathering co-occurrence statistics on the letters of a language to derive the probabilities of bigram and trigram sequences, it is possible to estimate the entropy at each point in a word. The method used by Juola, Hall and Boggs to estimate the entropy at a given point in the word was to consider the probability of all trigrams beginning with the last two characters encountered in the word. They then substituted as the third letter each character of the alphabet in turn, and using the formula

$$H = -\sum_{i=1}^{N} p(i) \log_2 p(i)$$

summed the log probabilities of each possible trigram to calculate the entropy. In this way they identified points in the word where entropy was higher than at neighbouring points, i.e., reached a local maximum. Their heuristic for morpheme identification was that if the entropy after any letter in the word was greater than the entropy after the previous letter, a morpheme boundary has been found. Of the 79 words tested by Juola, Hall and Boggs, 37 were completely correct, five were partially correct and 37 were wrong. The partially correct responses were cases in which either some but not all morpheme boundaries were correct, or where the morpheme boundaries were plausible but wrong, as for the word *noth-ing*, interpreted as the gerund of the pseudo-verb *to noth*.

4.5 Information theory and secrecy systems
Shannon believed that similar techniques should be used for machine translation as have been used for the decipherment of secret codes. He wrote that in the majority of secret code ciphers, it is only the existence of redundancy in the original messages that makes a solution possible (see Kahn 1966). From this, it follows that the lower the redundancy, the more difficult it is to crack a secret code. Low redundancy can be introduced into a message by using letters with more equal frequencies than would be found in normal text (such as partially suppressing the use of the letter *e* in English text), exercising greater freedom in combining letters and the suppression of frequencies by the use of homophones.

Reducing redundancy by techniques such as undoubling the sequence *ll* for Spanish messages will hinder the process of cryptanalysis. It follows that more text is needed to solve a low-redundancy coded message than one with a high-redundancy original text. Shannon quantified the amount of material needed to achieve a unique and unambiguous solution when the original message (or plain text) has a known degree of redundancy. He calls the number of letters **unicity distance** (see Denning 1982). One use of the unicity distance formula is in determining the validity of an alleged solution to a coded message or cryptogram. Shannon stated that if a proposed solution and key solves a cryptogram shorter than the unicity distance, then the solution is probably untrustworthy.

Shannon also wrote that a system of producing encrypted messages was like a noisy communication system. In information theory, noise is any unpredictable disturbance that creates transmission errors in any channel of communication, such as static on the radio, misprints, a bad connection on the telephone or mental preconceptions. It is also sometimes called thermal or white noise. Shannon states that the presence of noise means that the received signal is not always the same as that sent out by the transmitter. If a given transmitted signal always produces the same received signal, the effect is called distortion. However, the following discussion applies to cases where the same transmitted signal does not always produce the same received signal. If we call the entropy of the transmitted signal $H(x)$ and the entropy of the received signal $H(y)$, then $H(x)$ will be equal to $H(y)$ if there is no noise. We call the joint entropy of input and output $H(x,y)$, and also consider the conditional entropies $H_x(y)$ and $H_y(x)$. $H_x(y)$ is the entropy of the output when the input is known, and $H_y(x)$ is the entropy of the input when the output is known. All of these entropies can be measured on a per-second or a per-symbol basis, and are related by the following formula:

$$H(x, y) = H(x) + H_x(y) = H(y) + H_y(x)$$

Imagine a source which transmits a message consisting only of the digits 0 and 1, which occur with equal frequency. The noise in the system causes 1 digit in 100 to be received incorrectly. Thus if a 0 is received (and hence the output is known) the probability that a 0 was transmitted is 0.99, and the probability that a 1 was transmitted is 0.01. These values would be reversed if a 1 were received. These probabilities are conditional, since they depend on the knowledge of the output symbol. $H_y(x)$, the entropy of the input when the output is known, is then found by summing the products of the conditional probabilities and their logarithms. In this example,

$$H_y(x) = [0.99\log_2(0.99)+0.01\log_2(0.01)]=0.081 \text{ bits per symbol}$$

the conditional entropy $H_y(x)$ is called the **equivocation**, which measures the average ambiguity of the received signal. In the worst case, where a message consisting of equiprobable 0s and 1s is so affected by noise that the outputs are virtually random, the equivocation will be 1 bit per symbol.

4.6 Information theory and stylistics

Kondratov (1969) describes the application of information theory to poetics, and in particular studied the entropy of Russian speech rhythms. Stress in Russian may fall on any syllable of a content word which has mandatory stress, while non-content words may be joined to the content words to form a single rhythmic word. Thus, monosyllables mainly merge with neighbouring words.

The transcription method used by Kondratov to describe stress patterns in the spoken word was to use the number of syllables in a word as the base figure, and use a superscript figure to show the position of the stressed syllable. For example, 3^2 would denote a three-syllable word with the stress on the second syllable.

The frequencies with which each word type which occurred in selected texts drawn from five different genres was found, and were converted to entropy values

by applying Shannon's formula

$$H = -p_1 \log(p_1) - p_2 \log(p_2) - \ldots - p_n \log(p_n)$$

where n was the number of distinct word stress patterns encountered. Taking into account the number of words and syllables in each of the test texts the quantities entropy per word and entropy per syllable were also found. Scientific texts had the highest entropy per word, while the poems of Lomonosov had the least entropy both per word and per syllable. Kondratov concluded that the Russian language imposes constraints which influence the rhythmical regularity of speech, and these constraints must be greater for poetry than for other genres.

4.7 Use of mutual information in corpus linguistics

Church et al. (1991) state that mutual information (MI) can be used to identify a number of interesting linguistic phenomena, ranging from semantic relations of the doctor/nurse type (where the strength of association between two content words is found) to lexico-syntactic co-occurrence preferences between verbs and prepositions (where the strength of association between a content word and a function word is found). The higher the mutual information, the more genuine the association between two words. A table of mutual information values can be used as an index to a concordance. Church et al. assert that 'MI can help us decide what to look for in a concordance; it provides a summary of what company our words keep'.

They compared MI for the words *strong* and *powerful* with each of the words in the Associated Press (AP) corpus.[4] It was decided to compare the collocations of this pair of words because their meanings are so similar. Dictionaries are generally very good at identifying related words, but it is a more difficult task to describe exactly the subtle distinctions among related words such as near synonyms. The aim of this experiment was to describe the difference between two similar words in terms of their statistical behaviour. The top three scoring collocations in each case are shown in Table 2.2.

MI	Word pair	MI	Word pair
10.47	strong northerly	8.66	powerful legacy
9.76	strong showings	8.58	powerful tool
9.30	strong believer	8.35	powerful storms

Table 2.2 Highest scoring collocations of *strong* and *powerful*

The fact that people prefer saying *drink strong tea* to *powerful tea*, and prefer saying *drive a powerful car* to *a strong* car cannot be accounted for on pure syntactic or semantic grounds. Such lexical relations are called **idiosyncratic collocations** and account for many English word combinations. Another example is that *strong support* is more plausible than *powerful support*. In order to prove this, we could say that *powerful support* is implausible if we could establish that MI (*powerful, support*) is significantly less than zero. However, we are unlikely to observe MI scores much less than zero because these would only occur in an extremely large corpus. A more sensitive test for examining the dissimilarity between words is the *t* test, which is described in conjunction with the work of Church et al. in Chapter 1, Section 3.2.2.

As well as having applications in monolingual corpus linguistics, mutual information also has uses in multilingual corpora. For example, the automatic acquisition of bilingual technical terminology by Gaussier and Langé (1994) is described in Chapter 4, Section 3.2.6. Luk (1994) proposes the use of mutual information for the automatic segmentation of corpora of Chinese texts into their constituent words, and his method is described in Chapter 4, Section 3.2.11.

5 THE RELATIONSHIP BETWEEN ENTROPY, CHI-SQUARE AND THE MULTINOMIAL THEOREM

For the mathematically curious, we can now link the work from this chapter with that from the last. Consider a random variable which can take any one of *r* different values each with probability p_i, where *i* is in the range 1 to *r*, and the sum of all the p_i values is 1. For example, p_i could be the probability of each of the *r* categories of a linguistic form occurring in a segment of text. If this linguistic form is observed on *n* different occasions, and n_1, n_2, and so on up to n_r are the numbers of times each of the *r* categories is observed, then according to the multinomial theorem, the probability *P* of a particular distribution of categories is given by

$$P(n_1, n_2, \ldots, n_r) = \frac{n!}{n_1! \, n_2! \ldots n_r!} p_1^{n_1} p_2^{n_2} \ldots p_r^{n_r}$$

where $p_1 = n_1 / n, p_2 = n_2 / n$, and so on up to $p_r = n_r / n$.

According to Herdan (1962), the formula for entropy may be derived from the multinomial theorem using Stirling's approximation (see Mood, Graybill and Boes 1974). If the *observed* relative frequencies in a sample of a corpus n_i/n are denoted by p_i', and the relative frequencies in the corpus or language as a whole from the *theoretical* multinomial distribution are given by p_i, we have

$$\log_2 P = -n \sum p_i' \log_2(p_i') + n \sum p_i' \log_2(p_i)$$

The sum in each case is over all values of *i* from 1 to *r*. Herdan thus shows that the probability by the multinomial theorem is the sum of two different entropies, called the sample entropy $-n\sum p_i' \log_2(p_i')$

and the modified population entropy $+n\sum p_i' \log_2(p_i)$.

He also shows that entropy is related to the chi-square measure. According to the formula of von Mises (1931), when the number of observations is very great, the probability according to the multinomial theorem P is related to chi-square as follows:

$$P = Ce^{(-\text{chi-square}/2)} \text{ and thus}$$

$$\log P = \log C - \text{chi-square}/2$$

Combining this with the expression for log P as the sum of two entropies, we obtain

$$\text{chi-square}/2 = \log C + n\sum p_i' \log_2(p_i') - n\sum p_i' \log_2(p_i)$$

The distinction between P and C is as follows. P is the likelihood of obtaining the observed data given a theoretical distribution (such as obtaining 12 heads and eight tails from a fair coin), while C is the likelihood of obtaining the observed data given a distribution estimated from the observed data (such as obtaining 12 heads and eight tails from a coin which gives heads to tails in the ratio 12:8). If

$$P/C < e^{-3.84/2}$$

we can reject the null hypothesis (at the 5 per cent level) that the theoretical distribution and the actual distribution are equal.

6 SUMMARY

In this chapter, arguments were given in favour of probabilistic rather than rule-based models of natural language. In particular, it is simply not feasible to construct an all-encompassing set of rules to cover every conceivable feature of natural language. In addition, a yes/no rule states only that a linguistic feature is allowed, but does not say anything about the commonness or rarity of that feature. An important class of probabilistic models, namely Markov models, is based on Shannon's theory of communication. This theory is concerned with the amount of information generated by the source of a message. The term information does not refer to the semantic content of a message, but to the freedom of choice we have in constructing messages. The measure of information is called entropy. The greater the randomness or freedom of choice in a situation, and the fewer the constraints on that situation, the greater the entropy.

Markov models are probabilistic or stochastic processes, which may be depicted using graphs consisting of nodes corresponding to states of the model and vertices corresponding to transitions between the states. Each type of transition has an associated probability, and each time a transition between states is made, a symbol is emitted. In the simple Markov model, each succeeding state of the model depends only on the present state, but it is possible to construct more complex Markov models with memory, whereby previous states are taken into account when determining the probabilities of the next transition. Shannon has shown that Markov models can account for most but not all of the statistical dependencies observed in natural language. In an observable Markov model, any given state always causes the same symbol to be output, but for a hidden Markov model, the choice of output symbol is a probabilistic function of that state. Thus, there are three parameters defining a hidden Markov model: the set of transition probabilities, the set of output probabilities for each state of the system and the initial probabilities of the model. When hidden Markov models are used in real-world applications, three important problems that must be solved are the evaluation, estimation and training of the model parameters. The Cutting tagger is a part-of-speech tagger based on a hidden Markov model, where the Baum–Welch algorithm is used to estimate the initial model parameters. Both the Cuttting tagger and the CLAWS tagger use the Viterbi algorithm to estimate the most likely sequence of parts of speech given a sequence of lexical words.

Jelinek advocates the use of perplexity as an objective measure of language model quality. Proposed changes to a model can then be evaluated by examining whether they succeed in reducing the perplexity of the model. An optimal code is one which enables a message to be encoded using the fewest possible symbols, given a fixed alphabet.

Redundancy is the factor by which the average lengths of messages are increased due to intersymbol statistical behaviour caused by the statistical constraints within a language such as unequal symbol probabilities. The amount of entropy and redundancy in natural language can be estimated using Shannon's game. Information theory can be used for word segmentation, since morpheme boundaries tend to occur at points of maximum entropy. Information theory also has applications in secrecy systems and stylistics.

Closely related to information theory is the concept of mutual information, which is a measure of the degree of association between two events. It has been used by Gaussier and Langé to find the degree of association between pairs of technical terms in English and French found in parallel aligned corpora. Church et al. use mutual information for the identification of idiomatic collocations, while Luk uses this technique for the automatic segmentation of Chinese words.

In this chapter, therefore, we have seen what information theory and language modelling is, and how it has been used in corpus linguistics. By far the greatest contribution of information theory to date has been its ability to allow corpora to be part-of-speech tagged but this should not be allowed to overshadow those other potential uses for corpus linguistics that we have seen.

7 EXERCISES

1. In Shannon's game, one subject correctly guessed the letters shown in capitals at the first attempt, but guessed incorrectly the letters shown in lower case, as follows:

tHE shabBY reD cARPET

The spaces between words were counted as part of the test, and the subject guessed all of these correctly. Estimate the redundancy of English from this data.

2. a) Using the formula $H = -\sum p_i \log_2 p_i$, find the entropy of the word SHANNON. Notice that 1/7 of the characters are S, while 3/7 of the characters are N.

b) Calculate what the entropy would have been if the characters S, H, A, N and O were equiprobable.

c) Using the formula $H_{rel} = H/H_{max}$, calculate the relative entropy of the word SHANNON.

d) Using the formula $R = 1 - H_{rel}$, calculate the redundancy of the word SHANNON.

3. Consider the sentence *The stock market run continues.* Assume that *The* is always an article, *stock* could be a noun, verb or adjective, *market* could be a noun or a verb, *run* could also be a noun or a verb, and *continues* must be a verb. The matrix of transition probabilities derived from that used by CLAWS is as follows, where the first words are listed down the side and the second words are listed along the top:

	Noun	Verb	Prep.	Article	Adjective	Other
Noun	8	17	27	1	0	47
Verb	4	18	11	17	5	45
Prep.	22	4	0	46	6	22
Article	64	0	0	0	29	7
Adjective	72	0	5	0	3	20
Other	10	7	5	9	4	65

Using the Viterbi algorithm where the two most likely paths are retained at each step, find the most probable part of speech for each word in the sentence *The stock market run continues.*

8 FURTHER READING

Shannon's (1949) original account of his information theory is highly readable, and is recommended as a good introductory text. The application of information theory to linguistics is clearly and interestingly described by Paducheva (1963). Although Kahn's (1966) book is mainly about cryptography, it contains an interesting section about Shannon's work. It also contains a chapter called 'Ancestral Voices', which is relevant to the sections on decipherment in Chapter 5 of this book. Finally, *The Computational Analysis of English*, edited by

Garside, Leech and Sampson (1987), has several good chapters on the theory of CLAWS and probabilistic approaches to computational linguistics in general.

NOTES

1. Some have raised objections to this view. See Sperber and Wilson (1986).
2. Details of how to subscribe to this list can be found on the World-Wide Web at: http://www.hd.uib.no/fileserv.html . Alternatively, an e-mail message may be sent to the list administrator at: corpora-request@hd.uib.no .
3. The number of calculations required is $2TxN^T$ where T is the number of time steps in the model and N is the number of states in the model.
4. The Associated Press (AP) corpus consists of newswire reports in American English. The version used by this study consisted of 44.3 million words.

Clustering

1 INTRODUCTION

Willett (1988) describes **clustering** as the grouping of similar objects, and **cluster analysis** as a multivariate statistical technique that allows the production of categories by purely automatic means. Not only must similar objects be grouped, but dissimilar objects must remain distinct from each other. Two main types of clustering methods will described in this chapter; firstly, the related techniques of **principal components analysis** (PCA) and **factor analysis** (FA) which also group objects and variables; and secondly, cluster analysis techniques which require some calculation of the similarities between the objects to be clustered. Classification and categorisation are distinct concepts. Classification is the assignment of objects to predefined classes, while categorisation is the initial identification of these classes, and hence must take place before classification (Thompson and Thompson 1991). Categorisation is the task performed by the cluster analysis methods described in these sections. Anderberg (1973) states that automatic categorisation methods can be particularly valuable in cases where the outcome is something like 'Eureka! I never would have thought to associate X with A, but once you do the solution is obvious'. It is also clearly true that an algorithm can apply a principle of grouping in a large problem more consistently than a human can.

Clustering methods are said to be stable if the clusters do not change greatly when new data items are added; if small errors in the object descriptions lead to small changes in the clustering; and if the final categorisation does not depend on the order with which the data objects were encountered. Sneath and Sokal (1973) describe how cluster analysis was originally used mainly in the biological sciences, where it is referred to as numerical taxonomy. Nowadays, however, it is now used in many fields, including corpus linguistics, to perform such tasks as constructing a typology of English text types (Biber and Finegan 1986).

Section 2 of this chapter will describe how FA and PCA can be used for the clustering of variables in linguistics and in Section 3 we will cover the clustering of documents and corpus texts. In Section 4 we will see how terms can be clustered using their co-occurrence characteristics and approximate string matching techniques. Texts can be clustered according to similarities between the terms they contain, as described in Section 5, or on information theoretic grounds for sublanguage identification as described in Section 6. Finally, in Section 7, the clustering of dialects and languages will be described.

Occasionally in this chapter certain non-linguistic examples will be used, where this enables a point to be made as clearly and as simply as possible. The reader should not worry about having no knowledge of a particular application area and should concentrate instead on the general mathematical principles behind the example.

2 REDUCING THE DIMENSIONALITY OF MULTIVARIATE DATA

Woods, Fletcher and Hughes (1986) describe principal components analysis (PCA) and factor analysis (FA) as examples of multivariate techniques, where a number of variables are observed for each experimental subject, and we wish to summarise the information in the complete set of variables without designating any of the variables beforehand as either **dependent** or **independent**. In most quantitative studies, the conditions that are varied by the experimenter are called independent variables, while those variables whose response is being measured are called dependent variables (Butler 1985a). PCA and FA are methods of reducing the dimensionality of the data. That is to say, we attempt to reduce multivariate data to a single score for each subject, but retain most of the information present in the original data. For example, in the field of economics the retail price index is a single variable which summarises the fluctuations in the price of a large number of commodities, and the Dow-Jones index summarises the current price of a large number of shares. The advantage of using these measures is that one can make general statements about such factors as inflation without having to describe all the individual price changes. The disadvantage is that certain key data items might be concealed by the use of an overall index. For example, the Dow-Jones index might rise while one's own shares fall. In the calculation of the retail price index, the prices of the items in a basket of commodities are recorded. The price of each item is multiplied by a preselected weight, then the weighted prices are all summed together. An example of such an index in corpus linguistics is Yule's (1939) index of diversity (described in Chapter 5, Section 2.3); a single value for a given text which summarises the occurrences of every distinct word in the text. Sometimes it is not possible to reduce multivariate data to a single score, but one can still reduce the number of variables on which each subject is scored without significant loss of information. PCA and FA are examples of clustering techniques where groups of variables are clustered into a smaller number of factors.

2.1 Principal components analysis

Imagine the original data is in the form $(X_{i1}, X_{i2}, \ldots, X_{ip})$, where the ith subject is scored according to each of p variables. According to Woods, Fletcher and Hughes (1986), PCA is a mathematical procedure for converting these p original variables into a new set of p variables denoted $(Y_{i1}, Y_{i2}, \ldots, Y_{ip})$ for the ith subject.

The end-point of a PCA is to calculate a set of numerical weights or coefficients. A subject's score on one of the new variables Y_k is then calculated by multiplying the score on each of the original variables by the appropriate coefficient and then summing the weighted scores. If Y_{ik} is the score of the ith subject on the new variable, Y_k, then $Y_{ik} = a_{k1}X_{i1} + a_{k2}X_{i2} + a_{k3}X_{i3} + \ldots + a_{kp}X_{ip}$. The set of coefficients $a_{k1}, a_{k2}, \ldots, a_{kp}$ used to calculate a subject's score on the new variable Y_k are called the coefficients of the kth principal component, and Y_{ik} is the score of the ith principal component. The coefficients of the principal components are calculated to fulfil the following criteria as tabulated by Woods, Fletcher and Hughes (1986, p. 276):

1. The total variance of the principal component scores of the n subjects in the sample is the same as the variance of their scores on the original variables, i.e. , $\text{VAR}(Y_1) + \text{VAR}(Y_2) + \ldots + \text{VAR}(Y_p) = \text{VAR}(X_1) + \text{VAR}(X_2) + \ldots + \text{VAR}(X_p)$.
2. The coefficients $a_{11}, a_{12}, \ldots, a_{1p}$, of the first principal component are chosen so that $\text{VAR}(Y_1)$ is as large as possible. In other words, Y_1 is constructed to explain as much as possible of the total variability in the original scores of the sample subjects.
3. The coefficients $a_{21}, a_{22}, \ldots, a_{2p}$, of the second principal component Y_2 are chosen so that Y_2 is uncorrelated with Y_1, and Y_2 explains as much as possible of the total variance remaining after Y_1 has been extracted.
4. The coefficients $a_{31}, a_{32}, \ldots, a_{3p}$, of the third principal component Y_3 are chosen so that Y_3 is uncorrelated with both Y_1 and Y_2, and explains as much as possible of the total variance remaining after Y_1 and Y_2 have been extracted. This process continues until the coordinates of all p principal components have been obtained.

A simple non-linguistic example will now be given which clearly explains the principle of PCA. Jolicoeur and Mossiman (1960) used PCA to analyse measurements of the length, height and width of the carapaces of turtles. Their original data consisted of three dimensions, with X_1, X_2 and X_3 being the length, height and width respectively of each turtle shell. In Table 3.1, the coefficients of principal components derived by Jolicoeur and Mossiman are given. From this data we can define the first principal component as 0.81(length) + 0.31(height) + 0.50(width). Thus, a turtle shell with a length of 30 units, a height of 20 units and a width of 10 units would have a score on the first principal component of $(30 \times 0.81) + (20 \times 0.31) + (10 \times 0.50) = 35.5$.

Original dimension X	Component Y		
	Y_1	Y_2	Y_3
length X_1	0.81	-0.55	-0.21
height X_2	0.31	0.10	0.95
width X_3	0.50	0.83	-0.25

Table 3.1 Coefficients of principal components from a PCA of measurements of turtle shells

The starting point for the PCA in this case was the variance–covariance matrix of measurements of a sample of turtle shells, as shown in Table 3.2. The values on the main diagonal (top left to bottom right) are the variances of the corresponding variables, while the other terms are covariances. The matrix is symmetrical, meaning for example that the covariance between X_1 and X_2 (which is the degree to which they vary together) is the same as that between X_2 and X_1. Covariance (COV) and variance (VAR) values can be calculated from raw scores using the formulae

$$COV(X,Y) = \frac{1}{n-1}\sum(X_i - \overline{X})(Y_i - \overline{Y})$$

and

$$VAR(X) = \frac{1}{n-1}\sum(X_i - \overline{X})^2$$

where n is the number of experimental subjects, X_i and Y_i are the scores on the variables of interest for each subject, and $\overline{X}$ and are $\overline{Y}$ the mean scores over all subjects for the two variables of interest.

	length X_1	height X_2	width X_3
X_1	451.39	168.70	271.17
X_2	168.70	66.65	103.29
X_3	271.17	103.29	171.73

Table 3.2 Variance–covariance matrix of measurements of a sample of turtle shells

From Table 3.2 we see that the total variance is 451.39 + 66.65 + 171.73 = 689.77. PCA gives a first principal component Y_1 with variance 680.40, which is 98.6 per cent of the total variance. Thus, almost all the variability in the three dimensions of the turtle shells can be expressed in a single dimension defined by the first principal component. The coefficients of the principal components were calculated from the values of the variance–covariance matrix rather than the actual measurements of the shells.

The process of **reification** is to interpret each component meaningfully. In the turtles example, the first component is a size component, since all three dimensions are weighted positively. The other two components are components of shape, where a shell will score highly on the second component if it is short, high and wide, and highly on the third component if it is short, high and narrow.

In the context of language, PCA might be used for language test scores. A group of subjects might be scored on a battery of language tests, where the subtests measure different abilities such as vocabulary, grammar or reading comprehension. Woods, Fletcher and Hughes (1986, p. 279) describe the purpose of PCA as then being 'to determine how many distinct abilities (appearing as components) are in fact being measured by these tests, what these abilities are, and what contribution each test makes to the measurement of each ability'.

Alt (1990) gives another simple non–linguistic example of the correlation between hours of sunshine and temperature in towns on the south coast of England being found to be 0.9. The correlation coefficient between these two variables can be represented pictorially by an angle of approximately 25 degrees between the two variables when they are represented by vectors, since the cosine of 25 degrees is approximately equal to 0.9. The purpose of PCA might then be to answer the question posed by Alt (1990, p. 50). 'Can these two vectors be replaced by a single reference vector, known as a factor, such that the factor retains most of the information concerning the correlation between the original two variables?'

In order to summarise the two vectors (T representing temperature and S representing sunshine), one can bisect the angle between them to produce a reference vector called F_1. The reference vector makes an angle of 12.5 degrees with both T and S. The cosine of 12.5 degrees describes the correlation between the reference vector and T and S, and is 0.976. The correlation between a variable and a factor is called the **loading** of the variable on the factor. The square of the correlation coefficient between two variables, r^2, describes the amount of variance shared by these two variables, known as the **common factor variance**. The sum of the squares of the loadings of T and S on F_1, equal to 1.9, is known as the **extracted variance**. If T and S each has a total variance of one, as would be the case when using standardised z scores, the maximum variance that could be extracted by F_1 is $1 + 1 = 2$. The percentage of variance extracted by F_1 is therefore equal to $1.9/2 = 95$ per cent, and thus F_1 is a good summary of T and S, giving 95 per cent of the picture of the relationship between them. To account for the remaining 5 per cent of the variance, we must draw another vector F_2 at right angles (**orthogonal**) to F_1, which ensures that there is no correlation between F_1 and F_2, since the cosine of a right angle is 0.

The angle between T and F_2 is 102.5 degrees, and the angle between T and F_1 is 77.5 degrees. The loading of T on F_2 is then the cosine of 102.5 degrees, which is -0.216. and the loading of S on F_2 is the cosine of 77.5 degrees, which is 0.216. The amount of variance extracted by F_2 is $(-0.216)^2 + (0.216)^2$ which is approximately 0.1. Since $0.1/2 = 5$ per cent, F_2 has extracted the variance which remained unextracted by F_2. We have thus completed PCA as summarised in Table 3.3.

Variables	Factor	
	1	**2**
T	0.976	-0.216
S	0.976	0.216
Extracted variance	1.9	0.1
% total variance	**95**	**5**

Table 3.3 Principal components analysis for temperature T and hours of sunshine S

The sum of the squares of the loadings of each variable on the factor (**extracted variance**) is referred to as the **latent root** or **eigenvalue** when using terms taken from matrix algebra.

Alt (1990) suggests that to get an idea of how vectors are resolved into factors in several dimensions one should imagine a half-open umbrella, where the radiating spokes of the umbrella represent vectors and the handle represents a factor. The cosine of the angle between a spoke (vector) and the handle (factor) is the loading of the vector on the factor, where the loading of a variable on a factor expresses the correlation between the variable and the factor. The factor (represented by the handle) will give a better description of some vectors (represented by spokes making small acute angles with the handle) than others (represented by spokes making larger acute angles with the handle). To obtain a better resolution of the spokes (vectors) making larger acute angles with the handle, we would need a second handle (or factor). This would correlate better with some but not all of the factors. A large number of factors would be required to resolve each of the spokes. This analogy is intended to emphasise the following three concepts (Alt 1990, p. 60):

1. a large number of vectors (variables) can be resolved into a smaller number of reference vectors known as factors
2. the loadings of variables on factors express the correlations between the variables on the factors
3. different variables will load differently on different factors.

Alt (1990, p. 61) goes on to say that

the idea behind PCA is to extract factors sequentially such that the first factor accounts for the maximum common factor variance across all the variables (i.e., more than any other factor that could be extracted), a second factor is then extracted which is at right angles to the first (orthogonal to it) such that it extracts the maximum amount of the common factor variance that remains. A third factor is constructed, again at right angles, until all the common factor variance has been extracted (or enough as makes no difference).

In each case, factors which lie at right angles to each other are uncorrelated, since the cosine of 90 degrees is 0.

2.2 Conversion of loadings to weights

There is no direct relationship between factor loadings and factor score weights. For example, if one variable has twice the loading of another on a particular factor, this does not mean that the first variable contributes exactly twice as much to the factor as the second variable. The method of inter-converting factor loadings and weights is not described here,[1] but in general variables with high loadings on a factor also tend to have high factor score weights on the same factor.

2.3 Deciding on the number of principal components and their interpretation

Many of the components extracted by PCA will produce little variation between the samples, with differences between them being due to random error. This enables us to eliminate about half of the components. Eastment and Krzanowski (1982) suggest that components should be successively added until the inclusion of further components adds more noise than useful information. If no computer program is available for the Eastment and Krzanowski technique, the following guideline is frequently used: if the original data has p dimensions, only those components which account for more than the fraction $1/p$ of the total variance should be retained.

The **communality** of a variable (the sum of all the common factor variance of the variable over all factors) is the variance it shares in common with the other variables. If the communality of a variable is below 0.3, it is generally felt that it does not contribute enough to be included in PCA. Another frequently used rule is that one should extract only factors which have latent roots greater than one. This rule works well when the number of variables is 20 to 50. It is also common that only those factor loadings above 0.3 or below −0.3 should be considered significant.

Sometimes it is necessary to identify components subjectively, as was done in the turtle shells experiment, and sometimes it is not possible to identify a component at all. The identification of components in PCA is an area in which more research is needed. One method of interpreting principal components is to examine the correlation between each component and every variable. If a component is highly correlated with a particular variable then the component contains nearly all the information about differences in the subjects expressed by their scores on that variable.

In Woods, Fletcher and Hughes's (1986) experiment on learners' subtests, the first component was highly correlated with all subtests, so the first component had to be general linguistic ability. In general, when using PCA, the first component corresponds with a general attribute such as overall size or intelligence, or whether the writer of a text in an English corpus is or is not a native speaker of English. Later components tend to pick out more interesting specific shapes or skills. For example, the third component was correlated highly with six different variables, all related to speaking ability.

The correlations between components and variables can be calculated using the formula

$$r_{ij} = \frac{a_{ij}\sqrt{v_i}}{s_j}$$

where v_i is the variance of the ith component, s_j is the standard deviation of the jth variable, r_{ij} is the correlation between the scores of the subjects on the ith principal component and the jth variable and a_{ij} is the coefficient of the jth variable on the ith principal component.

2.4 The covariance matrix and the correlation matrix
In the turtle shells experiment the principal components were extracted from the covariance matrix of the subjects' scores. Components can also be extracted from the correlation matrix of the subjects' scores, as is done when using SPSS. The apparent number and structure of the significant components in PCA can depend on which of these two matrices is used as the starting point for the analysis.

The correlation between two variables is the covariance of these same variables when they are standardised by conversion to z scores, done by calculating how many standard deviations a variable is above or below the mean. If the data consists of variables which are not measurable on a common scale, such as age, social class and relative vowel type usage, they should generally be standardised, whereas if the variables are all of the same type, such as a set of scores on different language tests, they should be analysed in their original form. For all variables standardised by conversion to z scores, the standard deviation is unity. Thus, if we commence the analysis with the correlation matrix, the correlation between a component and a variable is simply

$$r_{ij} = a_{ij}\sqrt{v_i}$$

2.5 Principal component analysis and spoken data: Horvath's study of vowel variation in Australian English
Horvath (1985) analysed speech samples of 177 Sydney speakers, to determine the relative occurrence of five different variants of each of five vowel sounds. Using this data, the speakers clustered according to such factors as gender, age, ethnicity and socio-economic class. This type of study is becoming possible in corpus linguistics due to the availability of spoken corpora such as the Spoken English corpus (SEC).

According to Horvath, PCA is used widely in the social sciences for structuring large data sets into interpretable patterns. It is normally used with from 10 to over 100 variables. Other multivariate analyses such as multidimensional scaling are normally used with fewer than 10 variables. No variable should be included without a good theoretical or practical reason.

The original input to Horvath's PCA program consisted of the number of instances recorded in a speech sample for each variant of five vowel variables for each of the 177 Sydney speakers in the sample. The vowel variants were accented (A), ethnic broad (EB), cultivated (C), general (G) and broad (B). An example of the scores (number of occurrences of each vowel variant in speech) for one subject is given in Table 3.4.

iy					ey				ow				ay					aw				
A	C	G	B	EB	A	C	G	B	EB	A	C	G	B	EB	A	C	G	B	EB	A	C	G	B	EB
0	1	9	10	0	0	2	10	8	0	0	8	5	7	0	0	0	7	13	0	0	2	13	5	0

Table 3.4 Number of occurrences in speech of each vowel variant provided by one Sydney speaker

In such a study, there should ideally be at least a 4:1 ratio between subjects and variables. The first output of the PCA program was the matrix of correlations between the variables. The principal components were then extracted, along with their **eigenvalues** (the amount of variance accounted for by each component) and the **component loadings** (how the variables correlate with the principal component). For each subject, a speaker component score (that is, be a listing of each individual speaker in the data sample along with a component score for that speaker on each of the major principal components (PC)) was calculated as shown in Table 3.5. Four significant principal components were found.

	Speaker 8	Speaker 17
PC 1	3.67	−2.37
PC 2	0.82	−1.78
PC 3	1.45	0.52
PC 4	−0.21	0.96

Table 3.5 Sample speaker component scores

The individual speakers can then be clustered into groups of similar speakers by plotting them on graphs where their scores on principal components (PC1 and PC2, for example) are the axes, such as the one shown in Figure 3.1.

Each retained principal component is plotted against the others to achieve this clustering, enabling one to see the similarities between the individual

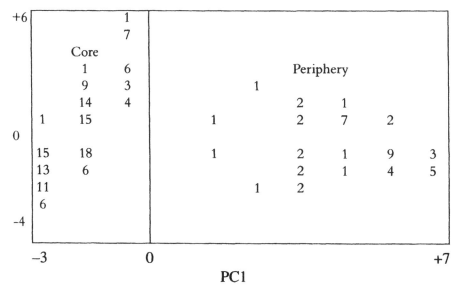

Figure 3.1 The Sydney speech community: core and periphery (PC1 x PC2)

speakers. Speakers whose linguistic behaviour is similar according to the criteria of the two principal components will plot closely together, and those dissimilar will plot further apart. The figures on the graph show the number of speakers who should be plotted at that point in the graph. Groups of speakers whose linguistic behaviour is similar can thus be identified, although this visual process is partially subjective. The interpretation of these graphs involved three steps; namely, defining linguistic groups, describing the linguistic characteristics of these groups and defining the social characteristics of those groups. The clusters of speakers identified in this way were characterised both by their linguistic behaviour and by their social make-up, and thus were referred to as **sociolects**. The signs of the loadings of the first principal component are shown in Table 3.6.

	accented	ethnic broad	cultivated	general	broad
iy	+	+	–	–	–
ow	+	+	–	–	–
ay	+	+	0	–	–
aw	+	+	–	–	–
ey	+	+	–	–	–

Table 3.6 Signs of the loadings for PC1 in Horvath's study of Sydney speakers

The first principal component split the Sydney speech community into two groups, which was clearly seen when PC1 was plotted against PC2. Those speakers with negative speaker-component scores formed the core of the speech community while those at the other end of the scale with positive scores were said to form the periphery. For speakers on the periphery, the C, G and B variants all loaded negatively, while the A and EB variants loaded positively.

A plot of PC2 against PC3 showed subdivisions within the periphery, according to gender and socio-economic class. For example, sociolects 1 and 2 consisted of males and lower-working-class speakers, sociolects 3 and 4 were mainly females and middle-class speakers. The members of sociolect 4 were all female, while sociolects 2 and 3 were mainly teenagers. Sociolect 1 had speakers of a predominantly English-speaking background and sociolect 4 had speakers from predominantly Greek-speaking backgrounds. The core speech community which used only the C, G and B variants of the vowels, can also be further subdivided. It can be demonstrated that linguistic change is taking place if older speakers tend to be in one group, while younger speakers are in another. This phenomenon is seen in the core speech community, as PC4 clearly separates teenagers and adults. A similar study clustered speakers according to variation in consonants, intonation and written texts obtained through a survey of childhood reminiscences. This study is a good example of 'eureka' clustering.

2.6 Factor analysis

Factor analysis (Spearman 1904) also determines the number of significant dimensions in a multivariate data set. For each extracted factor a set of loadings is obtained which are similar to the coefficients of principal components. In PCA, the variables of the principal components are assigned a unique set of variables, while in FA, through the process of **factor rotation**, the experimenter can choose a preferred solution from an infinite number of possible solutions. Examples of methods of rotation are the centroid method, principal-axes method, maximum likelihood method and Varimax. Rotating the factors does not change the total variance, but does alter the distribution of the variance across the factors. Rotation takes place so that loadings of moderate size found in PCA are either increased, or decreased to the point where they may be disregarded. The end result is that there are fewer variables that load significantly on a particular factor. Alt concludes that FA is not a pure mathematical procedure, since its success depends on an understanding of the phenomena being investigated. Owing to the complexity of the technique, computerised statistical packages such as SPSS are necessary to perform factor analysis.[2]

2.7 Factor analysis in corpus linguistics: the work of Biber and Finegan

Biber and Finegan (1986) and Biber (1988) describe their multi-feature multi-dimension (MFMD) approach to the examination of linguistic variation between text types. They use the term text type to describe the presence of linguistic features within a text, while the term genre is used to refer to the author's purpose in writing that text. The MFMD approach makes use of text corpora on computer (the Lancaster–Oslo/Bergen and London–Lund corpora) containing a wide range of texts and computer programs to count the frequency of linguistic features in each text. The potentially important linguistic features to be counted are chosen by a review of the research literature. Factor analysis is used to determine empirically which linguistic features tend to occur together in the texts. Finally, the text types are found using cluster analysis, grouping together texts which have the most similar linguistic features.

Having decided upon a set of potentially important linguistic features, the preliminary step to factor analysis is to calculate the correlation matrix of these variables based upon their patterns of co-occurrence. A method of converting co-occurrence data into correlation coefficients is given in Chapter 5, Section 3.1. For example, Biber (1988) identified 67 different linguistic features in a survey of the literature, then proceeded to derive a correlation matrix of those variables, from which factors could later be derived. The size of a correlation, positive or negative, shows to what degree two linguistic features vary together. Biber gives an example of a correlation matrix for four variables: first person pronouns (1PP), questions (Q), passives (PASS) and nominalisations (NOM). This hypothetical matrix is shown in Table 3.7.

	1PP	Q	PASS	NOM
1PP	1.00			
Q	0.85	1.00		
PASS	−0.15	−0.21	1.00	
NOM	0.08	−0.17	0.90	1.00

Table 3.7 Correlation matrix for the co-occurrence of four linguistic variables

Two pairs of variables are highly correlated, first person pronouns and questions, and passives and nominalisations. All the other correlations are much lower. This intuitively suggests that two distinct factors will be identified from this matrix, one consisting of first person pronouns and questions, and the other consisting of passives and nominalisations. These two factors are largely uncorrelated with one another, since the linguistic features on the first factor have low correlations with the features on the second factor. A factor analysis of this correlation matrix might produce the following two factors:

Factor A = .82(1PP) + .82(Q) + .11(NOM) − .23(PASS)
Factor B = −.16(1PP) −.19(Q) + .91(PASS) + .76(NOM)

The values assigned to each of the linguistic features on each factor are called **factor loadings**. These do not correspond exactly with correlation coefficients, but indicate a similar pattern.

Factor scores are used to examine the relations among the genres and text types with respect to each factor. A factor score is computed for each text by summing the number of occurrences of the features having significant weights of 0.35 or more on that factor, then normalising according to text length. A third factor identified by Biber and Finegan, designated reported versus immediate style, had positive weights greater than 0.35 for past tense, third person pronouns and perfect aspect. Those features with negative weights greater than −0.35 were present tense and adjectives. Thus, the factor score for factor 3 was found by adding together, for each text, the number of past tense forms, perfect aspect forms and third person pronouns (the features with positive weights) and subtracting the number of present tense forms and adjectives (the features with negative weights).

Genres can be compared by arranging them in order of their mean factor scores found by averaging the factor score of all the texts in that genre. For example, Biber and Finegan analysed various genres according to their factor scores on their first factor, interactive versus edited text. Their results are shown in Table 3.8, where genres such as telephone speech and interviews are close together near the interactive end of the scale, and academic prose and press reportage are close together at the edited end of the scale.

320	Telephone and face-to-face conversation
260	Interviews
200	Spontaneous speeches
180	Prepared speeches
150	Professional letters
120	Broadcast
100	Romantic fiction
90	General fiction
40	Belles lettres, hobbies
30	Official documents, academic prose
20	Press

Table 3.8 Factor scores obtained for various genres for interactive versus edited text

2.8 Factor analysis and a corpus of texts from the French Revolution

Geffroy et al. (1976) describe 'shifters' as being words which relate a text to the situation of communication. They used **factor analysis** to examine shifters in some texts of the French Revolution. Their first corpus of about 61,000 words comprised 10 speeches (denoted R1 to R10) made by Robespierre between July 1793 and July 1794. Their starting point for FA was a frequency table where n_{ij} was used to denote the frequency of word i in text j. In factor one *je* and *j'* were at the far end of the negative loadings. This shifter was found to be associated with R10, the last speech made by Robespierre, also known as *huit thermidor* where he speaks to defend himself against accusations made by his enemies, making frequent use of the pronoun *je*, suggesting a situation of isolation.

A second corpus of 34,500 words was made of writings by Hébert, comprising 23 issues of the revolutionary newspaper *Père Duchesne* from 1793, denoted H260 to H282. The most significant negatively-loaded feature of the first factor was the word *nous*. *Nous* acts as a shifter in the works of Hébert, because it occurs with both inclusive and exclusive meaning. Its inclusive form is seen in the first factor, since it co-occurs with *ils*, while its exclusive meaning is seen in the second factor, where *nous* and *vous* are both present. Hébert began to use the inclusive *nous* after the event called the *Fête de la Constitution* on 10 August 1793. This inclusive *nous* marks the union achieved between the *sans culottes* of Paris and those of the *départements*.

When the works of Hébert and Robespierre were combined for a joint factor analysis, factor one had negative loadings for *la, l', par* and positive loadings for *je, me, pas, j', ai, n', que, quand, on*. This factor allows these two authors to be distinguished, since Robespierre's speeches score negatively and all the issues of *Père Duchesne* score positively. Only Robespierre's final speech R10, which was characterised by a high use of *je*, was on the same side as the Hébert newspapers. Here FA was shown to be a means of differentiating between the styles of two different authors.

2.9 Mapping techniques

To represent the relationship between two variables one can draw a scatter plot. With three variables, interrelationships could be shown with a three-dimensional matchstick model, but there is no direct way of showing the relationships between four or more variables. For this we could examine scatter plots for all possible pairs of variables but this is not practical for large numbers of variables. It is possible to derive indirect representations of the inter-relationships between more than three variables using a variety of techniques called mapping or ordination methods. These seek to represent the original dimensionality of the data in a reduced number of dimensions, usually two, without significant loss of information (Alt 1990).

All mapping and clustering techniques depend on a measure of how similar objects and variables, for example, are to each other. These techniques take as their starting points estimates of the similarities between each and every object or variable under investigation. Correlation coefficients provide indirect esti-mates of similarity, while judgements of object similarity made by respondents provide direct estimates of similarity. A variety of measures used to describe the similarity or difference (distance) between two measures will be described in Sections 3.2 to 3.5.

2.9.1 Principal coordinates analysis

Principal coordinates analysis is a variant of principal components analysis. The first principal component may be thought of as the best line of fit to all the other variables represented as vectors. The positions of all the variables can then be specified as points on the first principal component. This form of representation is best if the amount of variance accounted for by the first principal component is high. An improved representation is obtained by plotting the variables onto the two-dimensional plane defined by the first two principal components which are represented by two vectors (axes) at right angles to one another. This was done in the study by Horvath (1985) described in Section 2.5, and illustrated in Figure 3.1.

2.9.2 Multi-dimensional scaling

Multi-dimensional scaling (MDS) is also a technique for constructing a pictorial representation of the relationships implied by the values in a dissimilarity matrix. There are many different mathematical methods and computer packages for carrying out MDS. The latter generally allow for **non-metric scaling**, where the **rank order of dissimilarity** between two elements is considered rather than the **absolute magnitude**. Provided non-metric scaling is used, the different methods tend to give similar results.

The technique of MDS was used by Miller and Nicely (1955) who produced a dissimilarity matrix for 16 different phonemes, depending on the proportion of occasions on which two phonemes were confused in noisy conditions. Although the phonemes were described initially by five different articulatory features or dimensions (voicing, nasality, affrication, duration and place of articulation), Shepard (1972), using this data, found a two-dimensional scaling solution, where all the phonemes could be depicted on the axes of nasality and voicing, as shown in Figure 3.2. MDS is often carried out as an exploratory technique to see whether the data structure can be described by a two-dimensional solution (Woods, Fletcher and Hughes 1986).

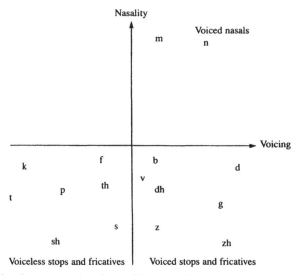

Figure 3.2 Multi-dimensional scaling of data on the perception of speech sounds

3 CLUSTER ANALYSIS

In the field of information retrieval, it is often necessary to sift quickly through a large number of documents in order to find and extract those relatively few documents which are of interest. This task is facilitated by **document indexing**, where each document is represented by a convenient number of index terms. Document indexing may be done automatically or by human indexers. Similarly, anyone wishing to view certain documents can express the topic of interest as a set of individual terms called **query terms**. Documents are retrieved whenever there is a sufficient degree of match between the query and document index terms. Similarly, in corpus linguistics, a query is often presented to a corpus, and sections of text (not necessarily whole documents) are retrieved in response. The main difference is that the corpus texts are indexed not by terms specially assigned to represent a much larger quantity of text, but by the words in the text themselves. Modes of corpus annotation can also act as retrieval keys, where, for example, we might search for all adverbs in a stated section of the corpus.

The information retrieval process may be enhanced by **document clustering**, where groups or clusters of documents which are relevant to each other are stored in similar locations, whether electronically or on bookshelves. Van Rijsbergen (1979) reports that the cluster hypothesis which underlies the use of document clustering is that closely associated documents tend to be relevant to the same requests. Thus, if one document is known to be of interest, the whole cluster will probably also be of interest. The following sections describe various methods of document clustering. Similar methods can be employed for the clustering of texts in a corpus, but with somewhat different aims. Rather than the resulting clusters merely being an aid to fulfilling future retrieval requests, in corpus linguistics the clusters are linguistically interesting in themselves. For example, sections of text written by a common author, or belonging to a similar genre, or being written in a common sublanguage, might be brought together. In corpus linguistics, not only sections of text, but the various identifiable linguistic features such as case, voice or choice of preposition within a text may be clustered. However, in standard information retrieval, the only two elements which are routinely considered are the documents and their index terms. Thus the two main types of clustering discussed in the information retrieval literature are document clustering and term clustering. With term clustering, each term in a document index can be replaced by an exemplar of the cluster containing that term. A cluster exemplar is a description of an average or most typical example of a cluster member. Queries made to a document database can be broadened by the addition of other members of the cluster.

3.1 Document clustering: the measurement of interdocument similarity

Document clustering analysis methods are all based in some way on measurements of the similarity between a pair of objects, these objects being either individual documents or clusters of documents (Willett 1988). To deter-mine the degree of similarity between a pair of objects, three steps are required:

1. the selection of the variables that are to be used to characterise the objects
2. the selection of a weighting scheme for these variables
3. the selection of a similarity coefficient to determine the degree of resemblance between the two sets of variables.

Citation clustering as described by Small and Sweeney (1985) involves measuring the degree of similarity between a pair of documents by the citations they share. However, in selecting the variables that are to be used to characterise the documents, it is more common to use document clustering techniques where documents are represented by lists of index terms, keywords or thesaurus terms that describe the content of the documents. With regard to the selection of a weighting scheme, Sneath and Sokal (1973) recommend that all the variables used in cluster analysis should be equally weighted. One reason is that in order to weight those attributes which are most important in determining the categorisation, such as attributes which rarely vary, we must know the categorisation in advance, so we cannot assign these a priori. Sneath and Sokal describe four main classes of coefficient for describing the degree of similarity between a pair of objects: distance coefficients, association coefficients, probabilistic coefficients and correlation coefficients.

3.2 Distance coefficients

According to Sneath and Sokal (1973), distance coefficients have intuitive appeal, and are often used to describe the evolutionary distance between two life forms or languages. Certainly distance measures are the most frequently applied. Distance is the complement of similarity, so two objects which are highly similar have little distance between them. In Euclidian distance, for example,

$$\Delta_{jk} = \sqrt{\sum_{i=1}^{n}(X_{ij} - X_{ik})^2}$$

the distance between two documents can be calculated as follows. Imagine that one document j is indexed by the terms *statistics, corpus* and *linguistics*, and another document k by the terms *corpus* and *linguistics*. Each document can now be represented by a sequence of 1s and 0s according to whether each index term in the list (*statistics, corpus, linguistics*) is present or absent. For document j, the first index term *statistics* is present, so $X_{1j} = 1$. The second and third index terms *corpus* and *linguistics* are also present for document j, so $X_{2j} = 1$ and $X_{3j} = 1$. For document k, the first index term *statistics* is absent, so $X_{1k} = 0$. However, the second and third index terms *corpus* and *linguistics* are present, so $X_{2k} = 1$ and $X_{3k} = 1$. Since we are considering a total of three index terms, $n = 3$. In order to solve the above equation, we must calculate $(X_{ij} - X_{ik})^2$ for every value of i in the range 1 to 3. $X_{1j} - X_{1k} = 1 - 0 = 1$, which multiplied by itself is 1. $X_{2j} - X_{2k} = 1 - 1 = 0$, which multiplied by itself is 0. Similarly, $X_{3j} - X_{3k} = 1 - 1 = 0$, the square of which is 0. Adding together the squares of the differences found for each value of i, we get $1 + 0 + 0 = 1$. Thus, the Euclidian distance between our pair of documents is the square root of 1 which is 1. Since Euclidian distance increases with the number of indexing terms used in the comparison, an average distance δ_{jk} is often calculated, as follows:

$$\delta_{jk} = \sqrt{\Delta_{jk}^2 / n}$$

The general form of another class of distance measures, the **Minkowski metrics**, is

$$d_r(j,k) = \left(\sum_{i=1}^{n} |X_{ij} - X_{ik}|^r \right)^{1/r}$$

The notation $|x|$ means that if x is negative, its sign must be changed to positive. If x is already positive, then it remains unchanged. If $r = 1$, the measure is called **Manhattan distance** or **City Block distance**. A variation of the Manhattan metric is the **Canberra metric**:

$$d_{CANB.}(j,k) = \sum_{i=1}^{n} \left(\frac{|X_{ij} - X_{ik}|}{(X_{ij} + X_{ik})} \right)$$

3.3 Association coefficients

Sneath and Sokal (1973) describe association coefficients as pair functions that measure the agreement in the attribute sets of two objects. Many of them measure the number of matches found in the two attribute sets as compared with the number of theoretically possible matches. According to Willett (1988), association coefficients have been widely used for document clustering. The simplest association coefficent is simply the value m, the number of terms common to a pair of documents. This measure has the disadvantage that it takes no account of the number of terms in each of the documents, and thus is not normalised. If we use the notation u to denote the number of unmatched terms, and n for the total number of index terms under consideration, then m can be normalised to form the simple matching coefficient:

$$S_{SM} = \frac{m}{m+u} = \frac{m}{n}$$

For example, when comparing the two sets of index terms (*carnation, lily,* rose) and (daffodil, lily, rose), the terms lily and rose match, so $m = 2$. The terms carnation and daffodil are not matched in each other's set of index terms, so $u = 2$. Thus

$$S_{SM} = \frac{2}{2+2} = 0.5$$

Association coefficients are most common with two-state characters, coded 0 and 1, such as the presence or absence of a linguistic feature or index term. All these measures are derived from the contingency table, a two-by-two table in which the cells are labelled a, b, c and d. Cell a records the number of times both attributes are positive, while cell b records the number of times the attribute is present in the first object but not the second. Cell c contains the number of times the attribute was absent in the first object but present in the second, and cell d records the number of times the attribute was absent in both cases. The number of matches $m = a + d$, while the number of mismatches $u = b + c$. Using the contingency table, the simple matching coefficient is

$$S_{SM} = \frac{a+d}{a+b+c+d}$$

Using the example of the two sets of index terms (carnation, lily, rose) and (daffodil, lily, rose), $a = 2$ (corresponding to lily and rose which occur in both sets of index terms), $b = 1$ (since carnation occurs in the first set of index terms but not the second), $c = 1$ (since daffodil is found in the second set of index terms but not the first), and $d = 0$ (since in this example we do not consider any index terms which occur in neither set). Thus:

$$S_{SM} = \frac{2+0}{2+1+1+0} = \frac{2}{4} = 0.5$$

which is the same result as before.

The coefficient of Jaccard is given by

$$S_j = a \ / \ (a + b + c),$$

and the Hamann coefficient by

$$S_h = (m - u) \ / \ n = (a + d - b - c) \ / \ (a + b + c + d)$$

The product–moment correlation coefficient r, described in Chapter 1, Section 4.3.1, and adapted for data coded 0 and 1 (denoting, for example, the presence or absence respectively of an index term) can also be calculated using the contingency table, and is given by

$$S_\theta = \frac{(ad - bc)}{\sqrt{(a+b)(a+c)(c+d)(b+d)}}$$

The wide variety of association coefficients described and tested by Daille (1995) are given in Chapter 4, Section 3.2.4.

3.4 Probabilistic similarity coefficients

Probabilistic similarity coefficients take into account the fact that a match between rare attributes such as individual open-class words is a less probable event than a match between commonly encountered attributes such as individual closed-class words in a corpus, and should thus be weighted more heavily. Examples of probabilistic similarity coefficients are joint information $I(h,i)$ which can be calculated from the contingency table using the formula $I(h,i) = n \ln n - a \ln a - b \ln b - c \ln c - d \ln d$ (where $\ln$ represents $\log_e$) and mutual information. Both these measures are described in detail in Chapter 2, Section 2.7.

3.5 Correlation coefficients

According to Sneath and Sokal (1973), one of the most frequently employed coefficients of similarity in numerical taxonomy is the Pearson product–moment correlation coefficient calculated between pairs of attribute vectors. It may be used on data where most of the attributes can exist in more than two states. In order to compute this coefficient between attribute sets j and k, the following formula may be used:

$$r_{jk} = \frac{\sum_{i=1}^{n}(X_{ij} - \overline{X}_j)(X_{ik} - \overline{X}_k)}{\sqrt{\sum_{i=1}^{n}(X_{ij} - \overline{X}_j)^2 \sum_{i=1}^{n}(X_{ik} - \overline{X}_j)^2}}$$

where X_{ij} is the attribute state value of attribute i in attribute set j, $\overline{X}_j$ is the mean of all state values for attribute vector j, and n is the number of attributes taken into account. This measure is related to the cosine measure given by Salton and McGill (1983), which is used to measure the similarity between a document and a user's query, both represented by a set of index terms, and returning a value in the range -1 to 1:

$$COSINE(DOC_i, QUERY_j) = \frac{\sum_{k=1}^{t}(TERM_{ik} \times QTERM_{jk})}{\sqrt{\sum_{k=1}^{t}(TERM_{ik})^2 \times \sum_{k=1}^{t}(QTERM_{jk})^2}}$$

$QTERM_{jk}$ represents the weight or importance of term k assigned to query j. $TERM_{ik}$ represents the weight of term k assigned to document i. The total number of index terms recognised by the matching system is t. If the two sets of index terms are considered as vectors in multi-dimensional space, the coefficient measures the cosine of the angle between documents or between queries and documents. The similarity between two vectors is inversely proportional to the angle between them, which is 0 if they are identical.

Sneath and Sokal suggest that one should use the simplest type of coefficient that seems appropriate. Such coefficients are often monotonic with more complex measures, meaning that they provide the same rank ordering (vary up and down together).

3.6 Non-hierarchic clustering

The two main classes of document clustering methods are non-hierarchic and hierarchic. With non-hierarchic clustering the data set is partitioned into clusters of similar objects with no hierarchic relationship between the clusters (Willett 1988). Clusters can be represented by their centroid, which could be seen as the 'average' of all the cluster members, and is sometimes called a **class exemplar**. The similarity of the objects being clustered to each cluster centroid is measured by a matching function or similarity coefficient. Non-hierarchical clustering algorithms use a number of user-defined parameters such as:

1. the number of clusters desired
2. the minimum and maximum size for each cluster
3. the vigilance parameter: a threshold value on the matching function, above which an object will not be included in a cluster
4. control of the degree of overlap between clusters.

The number of clusters may be fixed beforehand or they may arise as part of the clustering procedure (van Rijsbergen 1979). Non-hierarchical algorithms can be transformed into hierarchical algorithms by using the clusters obtained at one level as the objects to be classified at the next level, thus producing a hierarchy of clusters. In one version of a single-pass algorithm, the following steps are performed:

1. the objects to be clustered are processed one by one
2. the first object description becomes the centroid of the first cluster (van Rijsbergen 1979)
3. each subsequent object is matched against all cluster representatives (exemplars) existing at its processing time
4. a given object is assigned to one cluster (or more if overlap is allowed) according to some condition on the matching function
5. when an object is assigned to a cluster the representative for that cluster is recomputed
6. if an object fails a certain test (such as not matching any existing cluster sufficiently closely) it becomes the cluster representative of a new cluster (van Rijsbergen 1979, p. 52).

In the above procedure the first object description becomes the centroid of the first cluster, and thus we say that we have a single **seed point**, a seed point being whatever starts off a new cluster. There are a number of means whereby a set of k seed points can be used as cluster nuclei. The simplest method is to choose the first k units in the data set, but one must be sure that the data set is not already clustered. Alternatively one can subjectively choose any k units from the data set (so the data points are well separated) or use random numbers. Finally, if there are m objects to be clustered, the seed points can be spread throughout the sample by using object numbers m/k, $2m/k$, $3m/k$, and so on.

One example of a two-pass algorithm is MacQueen's (1967) k-means method. MacQueen's algorithm for sorting m data units into k clusters is composed of the following steps (Anderberg 1973):

1. take the first k data units in the data set as clusters of one member each
2. assign each of the remaining $m-k$ data units to the cluster with the nearest centroid. After each assignment, recompute the centroid of the gaining cluster
3. after all data units have been assigned in step 2, take the existing cluster centroids as fixed seed points and make one more pass through the data set assigning each data unit to the nearest seed point.

This clustering process can be further refined by reassignment of data objects into more suitable clusters. Each data unit should be taken in sequence and one should compute the distances to all cluster centroids. If the nearest centroid is not that of the data unit's present cluster, then the data unit should be reassigned to the cluster of the nearest centroid and the centroids of both the original and new clusters should be updated. This process should be repeated until processing all the items in the full data set causes no further changes in cluster membership.

3.7 Hierarchic clustering methods

According to Willett (1988), hierarchical document clustering methods produce tree-like categorisations where small clusters of highly similar documents are included within much larger clusters of less similar documents. The individual documents are represented by the leaves of the tree while the root of the tree represents the fact that all the documents ultimately combine within one main cluster. Hierarchical clustering methods may be either agglomerative (the most commonly used type of clustering procedure) or divisive.

With an agglomerative strategy, one starts with the individual documents, then fuses the most similar pair to form a single cluster. The next most similar document pair or document–cluster pair is fused, and so on until the entire document set forms a single cluster. This means that if n documents are to be clustered, a total of $n-1$ fusions must take place before all the documents are grouped into a single large cluster. With a divisive clustering strategy, we start with the overall cluster containing all the documents, and sequentially subdivide it until we are left with the individual documents. Divisive methods tend to produce **monothetic categorisations**, where all the documents in a given cluster must contain certain index terms if they are to belong to that cluster. Agglomerative methods tend to produce **polythetic categorisations**, which are more useful in document retrieval. In a polythetic categorisation, documents are placed in a cluster with the greatest number of index terms in common, but there is no single index term which is a prerequisite for cluster membership. Clearly both types of categorisation are relevant to corpus analysis.

3.8 Types of hierarchical agglomerative clustering techniques

There are several hierarchical agglomerative clustering methods, including the single linkage, complete linkage, group average and Ward (1963) methods. All these methods start from a matrix containing the similarity value between every pair of documents in the test collection. Willett (1988, p. 581) gives the following algorithm which covers all the various hierarchical agglomerative clustering methods:

> For every document (cluster) pair find $SIM[i\ j]$, the entry in the similarity matrix,
> then repeat the following:
>> Search the similarity matrix to identify the most similar remaining pair of clusters;
>> Fuse this pair K and L to form a new cluster KL;
>> Update SIM by calculating the similarity between the new cluster and each of the remaining clusters
> until there is only one cluster left.

The methods vary in the choice of similarity metric and in the method of updating the similarity matrix. For example, in the average linkage method, when two items are fused, the similarity matrix is updated by averaging the similarities to every other document.[3] A worked example is given in Table 3.9 overleaf.

In Table 3.9(a) the original similarity matrix showing the similarity coefficients between each of five documents is shown. The greatest similarity of 0.9 is found between documents a and c, so these are merged to form a single node ac. The matrix is then updated to form the one shown in Table 3.9(b). The similarities between the new cluster ac and each of the other documents are found by taking the average of the similarity between a and the document and the similarity between c and the document. For example, the new similarity value for ac and b is the average of the similarity value for a and b and that for c and b. In Table 3.9(b) the greatest similarity value is 0.8, between documents b and d, so these two nodes are merged to form a cluster, and the matrix is updated to produce Table 3.9(c). The greatest similarity value in this new matrix is 0.475, between the clusters ac and bd. These clusters are merged to form cluster $abcd$, and we obtain the final similarity matrix shown in Table 3.9(d). There is only one similarity value in the matrix, between cluster $abcd$ and document e. Once these are merged, all the documents are included within a single cluster.

	a	b	c	d	e
a	x	0.5	0.9	0.4	0.2
b	–	x	0.6	0.8	0.3
c	–	–	x	0.4	0.1
d	–	–	–	x	0.2
e	–	–	–	–	x

(a) Prior to clustering

	ac	b	d	e
ac	x	0.55	0.4	0.15
b	–	x	0.8	0.3
d	–	–	x	0.2
e	–	–	–	x

(b) After merging of nodes a and c

	ac	bd	e
ac	x	0.475	0.15
bd	–	x	0.25
e	–	–	x

(c) After merging of nodes b and d

	$abcd$	e
$abcd$	x	0.2
e	–	x

(d) After merging of nodes ab and cd

Table 3.9 Clustering of documents using the average linkage method starting with a similarity matrix

The results of this clustering process can be represented by a tree-diagram or **dendrogram**. To construct this dendrogram, draw vertical lines upwards from each node or document, then connect these vertical lines by a horizontal line at the point of similarity at the time the nodes are merged. The dendrogram for the documents discussed in Table 3.9 is given in Figure 3.3 below.

The single linkage approach differs from average linkage in the way the similarity matrix is updated after each merging of a node pair. At each stage, after clusters p and q have been merged, the similarity between the new cluster (labelled t) and another cluster r is determined as follows: If Sij is a similarity measure, denoting the similarity between nodes i and j, then $Str = \max(Spr, Sqr)$. Thus Str is the similarity between the two most similar documents in clusters t and r. The method is known as **single linkage** because clusters are joined at each stage by the single shortest or strongest link between them

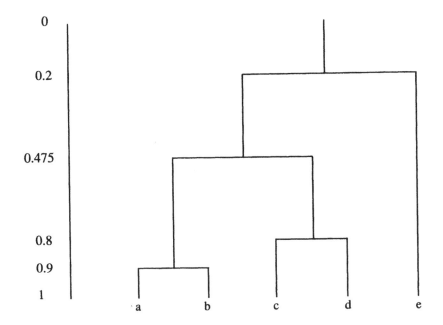

Figure 3.3 Dendrogram for the clustering of five documents

(Anderberg 1973). Single linkage is the best known of the agglomerative clustering methods, and is also referred to as nearest-neighbour clustering, since any cluster member is more similar to at least one member of its own cluster than to any member of another cluster. The method tends to produce long chain-like clusters, which can be a problem if nodes at opposite ends of the chain are greatly dissimilar, since ideal clusters should only contain objects which are similar to one another.

Complete linkage, or furthest-neighbour clustering, differs from single linkage in that the similarity between clusters is calculated on the basis of the least similar pair of documents, one from each cluster. At each stage, after clusters p and q have been merged, the similarity between the new cluster (labelled t) and another cluster r is given by $Str = \min(Spr, Sqr)$. The method is called complete linkage because all the documents in a cluster are linked to each other at a minimum level of similarity. The resulting clusters are a large number of small, tightly-bound groupings.

Ward's (1963) method joins together those two clusters whose fusion results in the least increase in the sum of the distances from each document to the centroid of its cluster.

The problem with using a similarity matrix as shown in Table 3.9 is that it becomes excessively large when significant numbers of documents are to be clustered. Most corpora are composed of a large number of documents, and so this is a genuine concern. As a solution to this problem, Anderberg (1973) describes the stored data approach, which differs from the stored matrix approach, in that similarity values are computed when needed rather than retrieved from storage. The steps are as follows:

1. for each row of the similarity matrix compute each similarity value, save the minimum/maximum entry in the row and record the column in which this extreme value occurs
2. search the row minima/maxima for the most similar pair
3. update the representation for the cluster resulting from the merger of the most similar pair. For each row involved in the merger, recompute the similarity values for the row and find a new row minimum/ maximum.

Willett (1980) describes an inverted file algorithm for the efficient calculation of interdocument similarities, which also avoids the creation of a similarity matrix.

3.9 The validity of document clustering

Clustering methods will find patterns even in random data, and thus it is important to determine the validity of any categorisation produced by a clustering method. One type of cluster validity study involves the use of distortion measures. These are quantitative measures of the degree to which the clustering method alters the interobject similarities found in the similarity matrix. The idea is to compare the results produced by different clustering methods on the same data. In general, methods which result in little modification of the original similarity data are considered superior to those which greatly distort the· interobject similarity data. The most common distortion measure is called the **cophenetic correlation coefficient**, produced by comparing the values in the original similarity matrix with the interobject similarities found in the resulting dendrogram. Conversely, Williams and Clifford (1971) suggest that the distortion of the similarity matrix is sometimes desirable because a clustering algorithm should try to find groupings that are more 'intense' than those present in the original similarity matrix. But whichever approach is adopted, clustering methods should be used with appropriate caution, and the issue of cluster validity should be considered in any study.

4 APPROXIMATE STRING MATCHING TECHNIQUES: CLUSTERING OF WORDS ACCORDING TO LEXICAL SIMILARITY

Hall and Dowling (1980) describe the use of **approximate string matching** (ASM) techniques for finding an item in a stored lexicon when there may be a spelling mistake or some other error in a keyword input by a user. These techniques include truncation (considering all terms commencing with a given number of identical letters as equivalent), stemming (determining the equivalence of terms through prefix and suffix deletion and substitution rules) and n-gram matching (determining the similarity of terms based on common character sequences). These can all be used as similarity coefficients for the clustering of terms according to their lexical similarity.

ASM techniques are employed by a number of automated intermediary systems, which guide non-expert users of bibliographic databases to select terms used in indexing documents. When ASM techniques are used in vocabulary selection, all the terms retrieved in response to a single input word may be considered members of a common cluster. The advantages of performing such term mapping are:

1. the user will discover which grammatical forms or orthographical variants (if any) of the terms of interest are in the lexicon
2. such mappings may retrieve hitherto unanticipated terms which describe the user's information need more succinctly
3. the mapping considers not only everyday English word variants, but can also consider domain specific prefixes and suffixes such as di-, tri-, methyl-, -ate or -ol, which indicate chemical structure.

User feedback should then determine which of the resulting cluster or list of proffered keywords truly reflect the user's information need (Oakes and Taylor 1991).

There are a number of areas of interest to us in which ASM techniques are used in the development of spelling checkers (see Sections 4.9.1 and 4.9.2), the automatic sentence alignment in parallel corpora where the same text is given in two different languages (see Section 4.9.3), and the identification of historical variants of modern words (see Section 4.9.4). ASM techniques enable words to be clustered according to their degree of lexical similarity to other words, but words can also be clustered on the basis of semantic similarity, as will be described later in this chapter. Section 5.1 covers Zernik's (1991) method of tagging word sense in a corpus, and in Section 5.2 Phillips (1989) uses the clustering of terms with related meaning to discover the lexical structure of science texts.

4.1 Equivalence and similarity

Both **equivalence** and **similarity** mappings are possible for words. If two character strings which are superficially different can be substituted for each other in all contexts without producing any difference of meaning, then they are said to be equivalent. Similarity is not necessarily transitive in this way; that is, if term A is similar to term B and term B is similar to term C, it does not necessarily follow that term A is similar to term C (Hall and Dowling 1980). Equivalence and similarity are key terms to consider in the course of the following pages.

Truncation and **stemming** directly generate equivalence classes or cliques, where each term indexed by that class will be retrieved by any member of that class. On the other hand, *n*-gram matching such as that described by Adamson and Boreham (1974) and word co-occurrence-based term retrieval employ a real-valued similarity metric for the pairwise comparison of terms. In such cases we must later employ thresholding to determine which lexicon terms should be output in response to a given input term. If terms are to be deemed equivalent on the basis of an above-threshold real-valued metric alone, overlapping clusters will be generated. For example, a term such as *bromopropylate* might be deemed equivalent to *chloropropylate* on the basis of the character structures of these terms, and *chloropropylate* might be considered equivalent to *chlorobenzilate*. But *bromopropylate* would not be considered equivalent to *chlorobenzilate* due to the transitive nature of the relation between these terms.

4.2 Term truncation

With simple truncation, an equivalence class consists of all terms beginning with the same *n* characters. For example, the English term *colour* and the French term *couleur* begin with the same two characters. Paice (1977) reports that the disadvantage of truncation is that there is no ideal value for *n*. For example, if *n* = 6, the words *interplanetary, interplay, interpolation* and *interpretation* will be incorrectly assigned to the same family, while the words *react, reacts, reaction, reacted, reactant* and *reactor* will all be assigned to separate word families. Truncation can often be effective, and is quite simple; consequently it is often included (for example, by Paice 1996) as a baseline against which to compare more complex term clustering methods.

4.3 Suffix and prefix deletion and substitution rules

Several sets of rules exist for the removal and replacement of common suffixes, including those produced by Lovins (1969), Porter (1980) and Paice (1977, 1990). Suffix and prefix removal assist in the retrieval of grammatical variants from a lexicon by generating equivalence classes or cliques, where each term indexed by that class will be retrieved by any member of that class. A second potential advantage of recognising suffixes and prefixes automatically is that these processes will aid the task of morphological analysis (described fully in Section 5.3) where (a) terms may be tagged according to the part of speech they represent on the basis of their suffixes, and (b) prefixes and suffixes often yield valuable domain-specific information about the activity of a term or the part of the body it pertains to. For example, the term *cardiopathy* has the prefix *cardio* meaning *heart* and the suffix *pathy* meaning *disease*, also denoting that *cardiopathy* is a noun.

The simplest form of stemming is the reduction of all terms to their inflectional root, which is the singular form for nouns and the infinitive form for verbs. This results in only a small proportion of equivalent terms in the lexicon being recognised, but does not result in any non-equivalent terms being confounded. This type of stemming has been described as **weak stemming** by Walker (1988). **Strong stemming** is the removal of longer and more meaningful suffixes such as *-isation, -ism* and *-ability*. Strong stemming often changes the meaning of the input term, as illustrated by the example *organ, organic, organism* and *organisation*. If one is to employ strong stemming, some form of human feedback is required to decide whether the stem is relevant or not. Stemming rules have also been produced for the French (Savoy 1993), Latin (Schinke, Greengrass, Robertson and Willett 1996), Malay (Ahmad, Yusoff and Sembok 1996) and Turkish (Kibaroglu and Kuru 1991) languages.

The rule sets for the removal and replacement of common suffixes cited above are given in the form of **production rules**. Production rules are defined by having both a condition and an associated action. Whenever the condition is true (in this case a particular suffix is identified), the action takes place (the suffix is removed or replaced, and the next rule is selected). Some of Porter's (1980) rules have more complex associated conditions, where rules may only become active if (a) the suffix is identified and (b) the suffix occurs in a given context. For example, the suffix *-ing* is removed only if the preceding consonant is undoubled. All these rule sets reduce terms to their 'derivational' roots, reducing, for example, the term *experimental* to *experiment*, where *experiment* is said to be the **base form** or **lemma** of the equivalence class. The entire rule set of Paice (1977) is reproduced in Table 3.10.

Starting with the rule at the top of the table, the endings in Column 2 are matched sequentially against the endings of the word being processed. When a match is found, the input word ending is removed and replaced by the ending shown in the third column. If the entry in the third column is null (denoted –), the ending is removed but not replaced. The process then terminates if the entry in Column 4 is *finish*, but otherwise continues from the rule indicated in that column. Paice states that it does not matter whether this process reduces the term to its linguistically correct root, provided (a) members of a word family are reduced to the same root, and (b) members of different word families are reduced to different roots. On this basis the lexical clustering of text can proceed.

Oakes (1994) describes a method of creating one's own domain-specific prefix and suffix rules. The first task is to produce an alphabetic list of domain terms, such as the *Derwent Drug File Thesaurus* (1986) which consists of pharmacology terms. One must then select the minimum root length, where only word pairs which match for this number of characters are initially considered to be in the same family. If two adjacent words match up to this root length, the terms should be matched up to the point of divergence, then both divergent endings stored. For example, if the minimum root length is four, the terms *followed* and *following* match beyond this point for two further characters.

Label	Ending	Replacement	Transfer
	-ably	–	go to IS
	-ibly	–	finish
	-ly	–	go to SS
SS	-ss	–ss	finish
	-ous	–	finish
	-ies	–y	go to ARY
	-s	–	go to E
	-ied	–y	go to ARY
	-ed	–	go to ABL
	-ing	–	go to ABL
E	-e	–	go to ABL
	-al	–	go to ION
ION	-ion	–	go to AT
	–	–	finish
ARY	-ary	–	finish
	-ability	–	go to IS
	-ibility	–	finish
	-ity	–	go to IV
	-ify	–	finish
	–	–	finish
ABL	-abl	–	go to IS
	-ibl	–	finish
IV	-iv	–	go to AT
AT	-at	–	go to IS
IS	-is	–	finish
	-ific	–	finish
	-olv	–olut	finish
	–	–	finish

Table 3.10 The term conflation rules of Paice (1977) for suffix replacement and removal

The distinct endings after the point of divergence are -*ed* and -*ing*, which are stored in a suffix frequency table. Once all adjacent word pairs in the lexicon have been considered, the frequency of each stored prefix is found. The list of most common suffixes, after manual post-editing to remove a proportion of nonsense suffixes, should yield a list of suffix-removal rules. To overcome the problem of overlapping suffixes, longer suffixes such as -*tic* should be placed in the table before shorter ones such as -*ic*. The same technique is used to create prefix-removal rules, except than one should begin with an alphabetic list of reversed lexicon terms. Using the *Derwent Drug File Thesaurus*, the following most common meaningful suffixes were identified: -*in*, -*al*, -*ine*, -*ium*, -*mycin*, -*ate*,

-amycin, -ic, -ol, -s, -an, -feron. The corresponding prefix list was: *a-, di-, an-, intra-, me-, para-, bi-, pro-, dia-, al-, micro-.*

Paice describes how prefix removal is more difficult than suffix removal, since the removal of a prefix will often radically alter the meaning of a word, as shown by consideration of the following two word families:

- *coordinate, inordinate, subordinate, ordinate*
- *bisect, dissect, insect, intersect, sect.*

4.4 Clustering of words according to the constituent *n*-grams

The use of stemming rules requires effort in the setting up of a suffix dictionary, and the use of look-up techniques during the processing of text. These problems may be overcome using the technique of Adamson and Boreham (1974). This is an automatic classification technique for words written in a given phonetic alphabet, based on the character structure of the words. In general, the character structure of a word may be described by its constituent *n*-grams, where an *n*-gram is a string of *n* consecutive letters all belonging to a standard alphabet (Yannakoudakis and Angelidakis 1988). Adamson and Boreham consider the number of matching bigrams (pairs of consecutive characters) in pairs of character strings, and employ Dice's (1945) similarity coefficient to cluster sets of terms from a chemical database. If a and b = total number of bigrams in the word A and B respectively, and c = number of bigrams common to A and B, then Dice's coefficient of similarity between A and $B = 2c/(a + b)$. For example, the words *pediatric* and *paediatric* may be divided into bigrams as follows: *pe-ed-di-ia-at-tr-ri-ic* and *pa-ae-ed-di-ia-at-tr-ri-ic*. The number of bigrams in each word is eight and nine respectively, giving a total of 17. The number of matching bigrams c is seven, so $2c/(a + b) = 14/17 = 0.82$. Dice's similarity coefficient is a real-valued similarity metric which has a value of zero if two terms are totally dissimilar in their character structures, one if two terms are identical or a value in-between if the character structures of the two terms partially match.

Input terms can be compared against each term in a corpus or lexicon to find those terms which have the greatest coefficient of similarity with respect to the user input term. Either the N most similar terms or all terms with an above-threshold coefficient of similarity are shown to the user, and if any of these are appropriate, they may be selected. Such a method can perform the function of prefix removal. It should also perform the function of a stemmer for suffix removal, with the advantage that it will cater for unanticipated suffixes, which would not be included in any suffix library. It should deal adequately with alternative spellings, such as Americanisation, as shown by the *paediatrics/pediatrics* example.

Angell, Freund and Willett (1983) describe a related method of comparing misspellings with dictionary terms based on the number of trigrams that the two strings have in common, using Dice's similarity coefficient as the measure of similarity. The misspelt word is replaced by the word in the dictionary which best matches the misspelling. They found that this resulted in the correct spelling being retrieved over 75 per cent of the time, provided that the correct version was in the dictionary. According to Mitton (1996), some trigram-based methods give more weight to some letters than to others. For example, letters at the front of a word might have more weight than letters at the end of a word, which in turn have more weight than letters in the middle of the word. In other systems, consonants may be given more weight than vowels. The system of Robertson and Willett (1992), described in Section 4.9.4, includes null characters at the start and end of words and thus effectively weights the characters at the front and end of the word more highly than the characters in the middle of the word.

Riseman and Hanson (1974) keep a record of all the trigrams which occur in the entire dictionary, and store these in a look-up table. To check the spelling of an input word, it is first divided into trigrams and each trigram is searched for in the table. If any input word trigram is found which does not occur in the table of dictionary trigrams, the input word is deemed to be a misspelling. This technique can detect errors made by an optical character reader, but is less appropriate for the identification of human spelling errors, as many of these, in particular real-word errors, do not include trigrams that never occur in the dictionary. Morris and Cherry (1975) divide input text into trigrams. They calculate an **index of peculiarity** for each word according to the rarity of its constituent trigrams. Mitton concludes that the advantage of n-gram based techniques is that they can be used for any language.

4.5 Longest common subsequence and longest common substring
Joseph and Wong (1979), in their work on the correction of misspelt medical terms, consider how many sections of the shorter string are present in the longer string. This is related to the problems of finding the longest common subsequence and the longest common substring.

Hirschberg (1983) defines the notions of a subsequence and a substring as follows: string a is a subsequence of string b if string a could be obtained by deleting zero or more symbols from string b. For example, *course* is a subsequence of *computer science*. A string c is a common subsequence of strings a and b if it is a subsequence of both. If string a can be obtained from string b by deleting a (possibly null) prefix and/or suffix of b, then we say that string a is a substring of string b. For example, *our* is a substring of *course*. A string c is a common substring of strings a and b if it is a substring of both. The main difference between a substring and a subsequence is that the substring consists of characters which must occur consecutively in the full string, while a subsequence consists of characters in the same order as they appear in the full string, although they need not appear consecutively in the full string. Algorithms for finding the longest common subsequence and longest common substring automatically have been given by McCreight (1976) and Wagner and Fischer (1974).

The notions of a longest common subsequence and a longest common substring are used in the definitions of various string to string dissimilarity measures. For example,

Coggins (1983) describes the following dissimilarity measures for clustering strings such as the character sequences of words:

$$d(a,b) = length(a) + length(b) - 2q(a,b)$$

where $d(a,b)$ is the dissimilarity measure for strings a and b, $length(a)$ and $length(b)$ are the lengths in characters of strings a and b respectively, and $q(a,b)$ denotes the length of the longest common subsequence between strings a and b.

$$D_s(a,b) = length(a) + length(b) - 2s(a,b)$$

where $s(a,b)$ is the length of the longest common substring between a and b.

$$D_c(a,b) = char(a) + char(b) - 2c(a,b)$$

where $char(a)$ gives the number of different characters from the alphabet used in string a, $char(b)$ gives the number of different characters from the alphabet used in string b, and $c(a,b)$ gives the number of different characters from the alphabet that x and y have in common. From the above three dissimilarity measures, the following composite measures may be derived:

$$D_2(a,b) = Ds(a,b) + d(a,b)$$

$$D_3(a,b) = (d(a,b) \times Ds(a,b)) + Dc(a,b)$$

Other weightings and combinations are possible.

4.6 Dynamic programming

Wagner and Fisher (1974) consider insertion of a character into a string, deletion of a character from a string and substitution (replacing one character of a string with another) in their dynamic programming system. In the unweighted case, each of these three operations is considered equally likely, and the distance between two strings is simply the least number of operations required to transform one string into another. For example, in order to transform the English word *colour* into the French word *couleur*, two operations are required: $c = c$, $o = o$, u deleted, $l = l$, e substituted by o, $u = u$, $r = r$. A more common measure is the **weighted Levenshtein distance**, where the cost of a substitution is two, while the cost of insertion or deletion remains at one. In the *colour/couleur* example we have one deletion and one substitution, so the weighted Levenshtein distance between them is three.

Kruskal (1983) describes the dynamic programming algorithm as follows:

Let m be the number of characters in the first string, and n be the number of characters in the second. Wagner and Fischer's matrix-filling algorithm computes d by constructing an $(m + 1)$ by $(n + 1)$ matrix, where the columns are labelled 0 to m and the rows are labelled 0 to n. The entry in cell (i,j) is the distance between the first i characters of word m and the first j characters of word n. The procedure is to calculate all the intermediate distances in the array, starting from the $(0,0)$ cell in the upper left-hand corner, and moving towards the (m,n) cell in the lower right-hand corner. The value in the $(0,0)$ cell is always 0, which is the distance between two empty sequences (nothing and nothing). The calculation of the values in the other cells then proceeds recursively as follows:

The values in all the other cells (i,j) are based on the values in the three predecessor cells $(i-1,j)$ $(i-1,j-1)$ and $(i,j-1)$. However, if $i = 0$ or $j = 0$, the two predecessor values involving negative values of i or j are not used. For each cell, three calculations are made:

- the value in cell $(i-1,j)$ plus the cost of deletion of the ith character in string a
- the value in cell $(i-1,j-1)$ plus the cost of substituting the ith character in string a (a_i) by the jth character in string b (b_j). The cost of substituting characters a_i and b_j is 0 if $a_i = b_j$.
- the value in cell $(i,j-1)$ plus the cost of insertion of the ith character in string a.

The value in cell (i,j) is the minimum of these three. In each case we add the cost of arriving at a state one operation away from transforming the first i characters of a into the first b characters of j plus the cost of that final operation. We cannot calculate the value in any cell before the values in the predecessors of that cell have been found, but otherwise the order of calculation is not important.

In the following example, illustrated in Table 3.11, the cost of transforming the word *cart* into *cot* is found using weighted Levenshtein distance, where the cost of insertion and deletion is one, and the cost of substitution is two if the substituted characters differ, and zero if a character is substituted for itself. For example, the entry shown in Column 2, Row one, shows the cost of transforming the first two characters of *cart* (*ca*) into the first one character of *cot*

		0	1	2	3	4
		–	c	a	r	t
0	–	0	1	2	3	4
1	c	1	0	1	2	3
2	o	2	1	2	3	4
3	t	3	2	3	4	3

Table 3.11 The cost of transforming *cart* into *cot*

(c), which is 1 for a single deletion. The entry in Column 4, Row 3 (the bottom right hand corner) shows the final result of the evaluation: the cost of transforming *cart* into *cot* is three.

Further extensions to the basic dynamic programming technique have been suggested. Lowrance and Wagner (1975) have extended this procedure to allow transposition of adjacent characters, by including in the minimisation function the quantity $d(i-2,j-2)$ + the cost of substituting with a_{i-1} with b_j + the cost of substituting b_{j-1} with a_j). Gale and Church (1993) in their work on sentence alignment, include three new operations in addition to those included by Wagner and Fischer. These are: expansion (one unit in the first string is replaced by two units in the second), contraction (two units in the first string are replaced by one in the second) and merging (two units in the first string are substituted for two in the second).

The dynamic programming procedure described above gives the alignment distance between two strings. If a record is kept of which operations were employed to arrive at each cell value by the use of pointers, then starting with the final cell (m,n) and using the pointer information repeatedly, we can obtain the actual path taken through the matrix all the way back to cell $(0,0)$. One potential use of this is that when comparing word pairs where one word comes from one language and one word from another, we can keep a record of exactly which letters were involved in the transformation process. In this way we can observe transformations which regularly occur between cognate words of a given language pair, such as the French *é* regularly being replaced by the English *s*.

4.7 SPEEDCOP

Pollock and Zamora (1984) describe SPEEDCOP (Spelling Error Detection/ Correction Project), an algorithm for correcting misspellings that contain a single error and whose correct forms are in a dictionary, as a coding method used for the identification of spelling errors in scientific databases produced by the chemical abstracts service. The codes are designed so that the codes of a misspelling and the corresponding correct word are identical or at least resemble each other more closely than do the original words (Robertson and Willett 1992). The SPEEDCOP system enables a measure of similarity between words. If the entire dictionary is transformed into SPEEDCOP codes and these codes are then sorted alphabetically, the similarity between a pair of words is the number of entries apart they are in this alphabetic list.

Two different types of code were developed, the **skeleton key** and the **omission key**. Pollock and Zamora describe how the skeleton key is constructed by concatenating the following features of the string:

- the first letter
- the remaining unique consonants in order of occurrence (thus doubled consonants must be undoubled where necessary)
- the unique vowels in order of occurrence (with the exception that if

the first letter is a vowel, it will appear again in the code if it appears later in the original word).

For example, the word *chemical* is encoded as *chmleia*. The rationale behind these coding measures is

- the initial letter is generally correct
- consonants carry more information than vowels,
- even in misspellings, the original consonant order is mostly preserved, and
- the code is unaffected by the doubling or undoubling of letters or most transpositions, features of typical spelling errors.

The main idea is that strings which appear similar have closely related keys, and thus subjective plausibility is more important than objectively measured similarity.

The main weakness with the skeleton key is that an incorrect consonant early in a word will cause it to be coded very differently to the desired word and thus to be sorted far apart in an alphabetic list of codes. To overcome this problem, the omission key was developed. Pollock and Zamora found that consonants were omitted from words in the following frequency order: *r s t n l c h d p g m f b y w v z x q k j* (thus *r* is omitted most often). The omission key for a word is constructed by sorting its unique consonants in the reverse of the above frequency order and then appending the vowels in their original order. Letter content is thus more important than letter order in the construction of the omission key. An example of the omission key is that *circumstantial* becomes *mclntsriua*.

The SPEEDCOP program corrects between 85 and 95 per cent of the misspellings for which it was designed. It also incorporates a common misspelling dictionary which contains keys to commonly misspelled words, and a function-word routine. Sometimes it is necessary to perform some form of ambiguity resolution to select between more than one potentially correct spelling of a misspelled word. For example, the input term *absorbe* might be intended to be *absorb* by deletion, *absorbed* by insertion or *absorbs* by substitution. Experiments showed that the order of precedence of these operations should be deletion = transposition > insertion > substitution, and thus the intended word is most probably *absorb*.

4.8 Soundex

The Soundex system which was originally developed by Margaret K. Odell and Robert C. Russell (see Knuth 1973) is also used as an aid to spelling correction. Both input words and dictionary words are again converted to codes, with the aim that a misspelled word should have the same Soundex code as the correct version of the word. The system has been used in conjunction with airline reservation systems and other applications involving surnames when these could be misspelled due to poor handwriting or voice transmission. The aim here is to transform all variants of a surname to a common code. The Soundex code is generated as follows (Knuth 1973, p. 392):

1. Retain the first letter of the name, and drop all occurrences of a,e,h,i,o,u,w,y in other positions.
2. Assign the following numbers to the remaining letters after the first:
   b,f,p,v : 1
   c g j k q s x z: 2
   d t: 3
   l: 4
   m n: 5
   r: 6
3. If two or more letters with the same code were adjacent in the original name (before step 1) omit all but the first.
4. Convert to the form 'letter digit digit digit' by adding terminal zeros (if there are less than three digits) or by dropping rightmost digits (if there are more than three).

Knuth gives the example that the names *Euler, Gauss, Hilbert, Knuth* and *Lloyd* have the codes E460, G200, H416, K530 and L300 respectively. The same codes will be obtained for the unrelated surnames of *Ellery, Ghosh, Heilbronn, Kant* and *Ladd*. Conversely, some related names like *Rogers* and *Rodgers*, or *Sinclair* and *St Clair*, are not transformed to identical Soundex codes. In general, however, the Soundex code greatly increases the chance of finding a surname given a close variant. Both the SPEEDCOP and Soundex systems have been used by Robertson and Willett (1992) for the identification of historical variants of words given their modern forms. These experiments are described in Section 4.9.4.

4.9 Corpus-based applications of approximate string matching

4.9.1 Typical spelling errors

The ASM techniques of truncation, stemming, *n*-gram, subsequence and substring matching give equal weight to all the characters considered in the analysis. The relative frequency of each character is not considered, nor is the fact that certain characters may be more commonly substituted for others within groups of spelling variants. The word-matching techniques of weighted dynamic programming and the SPEEDCOP and Soundex systems, on the other hand, all depend on prior analyses of the relative occurrences of actual spelling errors. In order to make such analyses, a number of authors have built corpora of real-life spelling errors.

Mitton (1996) describes a corpus of English spelling errors consisting of a collection of ten-minute compositions on the topic *Memories of my primary school*. This was designed to conform with the ideals of including a cross-section of adults rather than children and using free writing rather than spelling tests or psycholinguistic experiments such as the Cloze test.

He writes that real-word errors (such as *forth* for *fourth*) are the most difficult for spelling checkers to deal with, and are relatively common: about 40 per cent of the errors in his corpus were of this type. Real-word errors may be wrong-word errors, where some other word was written in place of the correct one.

The commonest wrong-word errors in the corpus were *too/to, were/where, of/off* and *their/there*. A second class of real-word errors involves the use of a wrong form of the intended word rather than the use of a different word altogether. The most common single error of this type was *use to* for *used to*. A third class of real-word errors were words incorrectly divided into two simpler words, such as *my self* or *in side*. Mitton found that in 93 per cent of cases, misspellings were correct in the first letter, and many of those cases in which the first letter was incorrect involved silent first letters as in *know* and *write*.

Slips and typos were described as errors that occur when people know how to spell a word but accidentally write or type something else. Damerau (1964) found that about 90 per cent of typing errors are single-error misspellings. Of these, the four most common types are insertion (one extra character inserted into the string), omission (one character removed from the string), transposition (where two adjacent characters in the string are interchanged) and substitution (one character in the string is replaced by another). According to Mitton, omission can occur when the key is not struck hard enough, especially when using the little finger; insertion occurs when one finger hits two keys, so the inserted letter is usually either a neighbouring character on the keyboard or an incorrect doubling of the character; substitutions are usually of neighbouring characters; and most transpositions involve keys typed with different hands, where, for example, *th* is often mistyped *ht*.

Yannakoudakis and Fawthrop (1983a) analysed the error patterns in a corpus of spelling errors consisting of 60,000 words of continuous text written by three adults who considered themselves to be very bad spellers. They found that the majority of English spelling errors are highly predictable and also that few errors are made in the first character of a word, but rather are made in the use of vowels and arise in the use of w, *y* and *h*. In relation to Damerau's four single-error types, they report that:

- doubling and singling of any letter is common, the most common instances being an incorrect doubling of *l* and an incorrect singling of *ss*. Singling was over twice as common as doubling
- transposition of any two adjacent characters is common
- certain consonants are more frequently interchanged than others. For example, they tabulate the number of instances in which combinations of *c, k* and *s* are interchanged with each other.

They also report that the type of spelling errors made depended on the regional dialect of the subject. All vowel errors were found to have some phonetic basis, such as *or* being substituted for *au*. Rules corresponding to all these error types were incorporated into a spelling correction system (Yannakoudakis and Fawthrop 1983b).

Mitton reports that the most common method employed by spelling checkers and correctors is simply to look up the input word in a dictionary. If the word is not there, it must be an error. The dictionary may be enhanced by allowing the user to build a private supplement to it – some systems store stems and rules for affix stripping, as was previously described in Section 4.3. Other systems use **hashing functions**, methods whereby words are converted to numbers and more than one word may have the same number or hash code. The system is trained by passing through a sufficient body of typical text, and a record is kept of all the hash numbers encountered. In the future, words hashing to numbers not previously encountered will be rejected.

The task of spelling correction is more advanced than mere spelling error detection. Spelling correction systems must not only point out spelling errors, but must aim to suggest the word the user intended to type. According to Mitton, commercial companies tend not to publish details of how their spelling correctors work. In sorting a list of suggestions, use may be made of a knowledge of word frequencies or context in which the word occurs, as described in Section 4.9.2.

4.9.2 Use of context to disambiguate real-word errors

As we have seen, a traditional dictionary-based spelling error detection program determines that a word is misspelled if it does not appear in the program's dictionary. Some misspellings, described by Mitton (1996) as real-word errors, give rise to another legitimate word in the dictionary, and thus are undetectable by this method. The number of possible undetectable errors is proportional to the number of words in the dictionary. For example, a small dictionary may not contain the word *chat*, and thus if the word *chat* were written in place of *that* the error would be noticed. However, a larger dictionary may well contain the word *chat*, causing this same error to pass unnoticed. One partial solution is to vary the words in the dictionary by subject area. An example of this is given by Atwell and Elliott (1987), who worked on a corpus of misspellings: *current* is much more likely than *currant* in the context of electronics. Another improvement to the basic dictionary method would be to store word frequency information with the dictionary, so the user could be informed if the selected word were uncommon and therefore possibly incorrect.

Mitton states that efforts to detect real-word errors generally depend on part-of-speech tagging the words in the context of their sentence and signalling a part-of-speech sequence with low probability, as exemplified by Atwell and Elliott's use of the CLAWS probabilistic tagger described in Chapter 2, Section 4.1.1. Mays, Damerau and Mercer (1991) use word co-occurrence trigrams rather than morphosyntactic tag bigram occurrence data to identify improbable word combinations. The main task of such context-based spelling correction is determining the degree of syntactic, semantic and pragmatic feasibility (represented by word trigram or tag bigram probabilities, gathered from statistical studies of a large corpus of text) of alternative sentences.

The system of Mays, Damerau and Mercer might consider a phrase taken from transcripts of the Canadian parliament which should read *I submit that what is happening in this case*. If the first word were to be misspelt, we might have *a submit that what is happening in this case*, and if the second word were misspelt we might have *I summit that what is happening in this case*. In order to determine which of these three phrases is the likeliest, the natural logarithms of word trigram probabilities are considered. Since logarithms are additive, the logarithm of the probability of the entire phrase can be found by summing the logarithms of the probabilities of each of the trigrams found within the phrase with two null symbols at the start. When the correct version of the phrase is divided into the trigrams *null null I* + *null I submit* + *I submit that* + ... + *in this case*, the sum of the logarithms of the trigram probabilities is −39.1. For the first error sentence this value is −43.2, and for the second error sentence it is −52.1. This shows that the correct sentence is the likeliest of the three possibilities, since it has the least negative probability. We can see this pattern emerging when we compare the logarithm of probability (log p) of *I submit that* (−1.2) with the log p of *I summit that* (−5.5) and the log p of *a submit that* (−3.7).

Atwell and Elliott use CLAWS unusual tag pairs. Since CLAWS makes no absolute distinction between ill-formed sentences and those which are correct but rare, a probability threshold must be employed below which input from a corpus is deemed incorrect. Whenever an incorrect word sequence is detected, it is possible that a real-word error has occurred, and Atwell and Elliott propose that 'cohorts' should be generated for each word in the sequence, where a cohort is a list of the input term and several other terms which might be the intended form of the input term.

The rule system of Yannakoudakis and Fawthrop (1983b) was designed to suggest corrections when input words are not found in the dictionary, but could also be used to generate cohorts for valid English words which are found in unlikely tag sequences, in the hope that one member of the cohort is more syntactically appropriate. For example, the incorrect input phrase *I am very hit* would generate the following cohorts, shown in Table 3.12.

I	am	very	hit
	an	vary	hot
	a	veery	hut
			hat

Table 3.12 Cohorts generated for the phrase *I am very hit*

Each member of a cohort should be assigned a relative probability value, taking into account the following factors:

- its degree of similarity to the word actually typed, where the typed word itself scores 1

- the trigram probability of the syntactic tags of each cohort member and the tags of the words immediately before and after
- the frequency of use of the word in English in general (Here *very* would get a high probability value while *veery* would get a low probability value)
- a high weighting should be given for preferred idioms like *fish and chips*, and a much lower weighting for a word sequence not in the dictionary of idioms such as *fish and chops*
- weighting should be given for domain dependent lexical preferences.

All these factors should be multiplied together to give the final probability weightings.

Cohorts could be stored with each dictionary entry to save calculating them each time. Since it would require considerable research effort to produce a lexicon for each domain and a lexicon of collocational preferences, Atwell and Elliott suggest a simpler system in which each dictionary entry holds (a) the word itself, (b) the syntactic tags normally associated with that word and (c) error tags, which are the tags associated with any similar words where these are different from the word's own tags. An error would be reported whenever an error tag is found to be more probable in the given context than any of the word's own set of possible tags. An even simpler method would be to store a matrix of the error likelihoods of tag pairs, based on the analysis of a large corpus of real-life errors.

4.9.3 Automatic sentence alignment in parallel corpora

In a parallel corpus,[4] the same body of text appears in two or more languages. The task of sentence alignment is to postulate exactly which sentence or sentences of one language correspond with which sentence or sentences of the other language. An aligned parallel corpus provides an aid to human translators since it is possible to look up all sentences in which a word or phrase occurs in one language and find exactly how these sentences were translated into the other language. In other words, it enables multilingual concordancing. Sentence alignment also facilitates word alignment. Various statistical measures (described in Chapter 4, Sections 3.2.4 and 3.2.5) exist which determine instances where a word in one language consistently appears in sentences which align with sentences containing a word in the other language. The Gale and Church (1993) sentence alignment method is based on the facts that

- longer sentences in one language tend to be translated into longer sequences in the other
- certain types of alignment are more commonly encountered than others.

All possible sentence alignments are considered by the dynamic programming technique which eventually finds the most likely sequence of alignments. As each putative alignment is considered, a penalty is given according to the empirically determined a priori likelihood of the alignment, as shown in Table 3.13. The most common alignment type is where one sentence of one language matches just one sentence of the other language. This is called a 1:1 alignment. Other possible alignment types considered by Gale and Church are 1:0 or 0:1, where a sentence is present in one language but is not translated into the other, 2:1 or 1:2 where two sentences of one language correspond with just one of the other, and a 2:2 'merge' which consists of a pair of sentences in each language, where neither sentence of the first language corresponds exactly with either sentence in the second, but both sentences taken together from the first language correspond exactly with both sentences taken together from the second language.

Alignment Type	1:1	1:0 or 0:1	2:1 or 1:2	2:2
Probability	0.89	0.0099	0.089	0.011
Penalty[a]	0	450	230	440

Table 3.13 Penalties for various alignment types according to Gale and Church (1993)

[a]Penalty = -100 * log [probability of the alignment type/probability of a 1:1 alignment]

A second penalty is added if the two sentences differ in their lengths in characters. An allowance is made for the fact that, for example, 100 characters of English on average correspond with about 120 in French or about 20 in Chinese. If we consider the bell-shaped curve which corresponds to the expected distribution of sentence lengths and draw vertical lines cutting the x-axis at the points corresponding to the observed sentence lengths, the penalty will be proportional to the area under the curve between the two lines. The most likely sequence of alignments is the one which incurs the least total penalty. In practice, the algorithm works well for 1:1 alignments, but the error rate is five times as great for 2:1 alignments, and 1:0 alignments are often totally missed (Simard, Foster and Isabelle 1992).

Simard, Foster and Isabelle found that a small amount of linguistic information was necessary in order to overcome the inherent weaknesses of the Gale and Church method. They proposed using **cognates** which are pairs of tokens of different languages which share obvious phonological or ortho-graphical and semantic properties. Their criteria for cognates were

- punctuation marks
- if the words contain digits, they must be identical
- if the words contain only alphabetic characters, the first four must be identical.

They thus have a binary classification where pairs of words are either cognates or not, but McEnery and Oakes (1996) use real-valued approximate string matching measures to estimate the extent to which words are cognate, namely truncation, Dice's similarity coefficient on the bigram structure of terms and dynamic programming. In comparing English and French vocabulary from a telecommunications corpus, it was found that cognates determined by bigram matching were accurate 97 per cent of the time if Dice's similarity coefficient was equal to or greater than 0.9, and 81 per cent if Dice's similarity coefficient was in the range 0.8 to 0.9. Accuracy of 97.5 per cent was found if truncation length was eight, falling to 68.5 per cent if truncation length was six. Dynamic programming was found to be slightly less accurate than Dice's similarity coefficient for bigrams.

4.9.4 Use of approximate string matching techniques to identify historical variants

Robertson and Willett (1992) discuss a range of techniques for matching words which occur in the Hartlib papers, a corpus of 17th-century English texts with their modern equivalents. They wanted a user to be able to retrieve from the 17th-century corpus using modern standard spellings. Although modern English spelling is largely standardised, in the 17th-century a given word might appear in any of several equally valid forms. A variety of spellings of the same word might occur within a single text. For example, letters would be added or deleted for line justification. Thus some modern words have more than one associated old form.

The experiments of Robertson and Willett involved the use of n-gram matching, SPEEDCOP coding and dynamic programming methods for spelling correction. In the n-gram experiments, both bigrams and trigrams were used. Their method of dividing a word into its n-grams differed slightly from the method described in Section 4.4, since one padding space was added to each end of every word before its division into bigrams, and two padding spaces were added to each end of the word before its division into trigrams. For example, the word *string* results in the generation of the bigrams *s, st, tr, ri, in, ng, g*, and the trigrams **s, *st, str, tri, rin, ing, ng*, g**, where * denotes a padding space. The method assumes that historical variants of words having the greatest numbers of n-grams in common with a given modern form are most likely to be associated with it. The method was deemed successful if a historical variant was one of the 20 historical words which best matched its modern form.

In the SPEEDCOP experiments, each of the words in the Hartlib test collection were sorted into two files, one consisting of each skeleton key and its original word, and the other of each omission key and its original word. Each of these files was sorted into alphabetical order of the respective key. The position of the key of each modern form considered by the experiment was found, and the coding method was deemed successful if the key of its historical variant was within 10 places of the key of the modern form. The dynamic programming experiments used the unweighted algorithm of Wagner and Fischer, described in Section 4.6. Once again, the 20 most similar historical words to the modern test word were retrieved, and the method was deemed successful whenever a historical variant of the modern word was among them.

Both of the n-gram methods were found to give high recall (that is, they retrieved a high percentage of the word variants in each case), achieving 95 per cent and 89 per cent for bigrams and trigrams respectively. The skeleton key outperformed the omission key, but produced only 76 per cent recall. Dynamic programming produced 96 per cent recall, but took about 30 times as long as bigram matching. Since Robertson and Willett were concerned with the retrieval of the 20 most similar old forms in each case, precision (percentage of words retrieved which were true variants of the input word) was inevitably low.

5 CLUSTERING OF TERMS ACCORDING TO SEMANTIC SIMILARITY

5.1 Zernik's method of tagging word sense in a corpus

Zernik's method (1991) identifies the set of senses for a polysemous word such as *train* (which, for example, can mean *to educate* or *railway train*). To summarise Zernik's method, the word under investigation is looked up in a corpus using a concordancer, and the lexical and semantic information contained within each concordance line is summarised by a set of weights called a **signature**. A hierarchical clustering algorithm is used to cluster the concordance lines according to their signatures, so that concordance lines containing a particular sense of the word of interest are brought together.

Each concordance line consists of five words before and five words after each occurrence of the word under investigation, and is represented as a list of weighted features such as terms or part of speech (the signature). A number of features are used as they are found in each line of the word concordance in the derivation of the signatures, including the appearance of

- a full word
- a word stem
- collocations (information about collocations is stored in the lexicon)
- part of speech.

Weight is a function of probability, so, for example, high-frequency terms have low weight, and a part of speech rarely taken by a word is given a high weight. The actual formula is $s(X) = \log(1/p(X))$ where $s(X)$ is the saliency or weight assigned to a feature and $p(X)$ is the probability of a feature occurring.

For example, out of 1.3 million words in the entire corpus, the word *rate* occurred 115 times. Thus, the probability of the word *rate* appearing within a ten-word window was 115*10/1.3 million. Also, out of the 4116 concordance lines for the lemma *train*, the word appears 365 times with the suffix *s*. Thus, the probability of train occurring with the suffix *s* is 365/4116 = 0.089. Each signature is first normalised by the sum of its weights. For example, if a signature consists of three features with raw weights 1, 2 and 3, the sum is 6 and the normalised weights are 1/6, 2/6 and 3/6. Hierarchical clustering is used for word-sense classification. A distance function between signatures is defined as the dot product of the two signatures. The distances are used in a complete-linkage clustering algorithm to produce a clustered hierarchy of concordance lines. As a result of this clustering, a binary tree is produced with all the concordance lines as leaves. Zernik's system was found to be 95 per cent accurate for *train*, where the two senses of the word take different parts of speech, but was totally unable to disambiguate *office* which is a noun in both its senses.

5.2 Phillips's experiment on the lexical structure of science texts

Phillips (1989) describes how items can be used as technical terms with quite distinct meanings in different areas of science – for example, *solution* has different meanings in mathematics and chemistry. Thus, we can compare vocabulary and genre in scientific texts, and collocational patterning will vary with genre.

In Phillips's experiments, a specialised corpus was used, drawn from the Birmingham University corpus of modern English text. This subcorpus of half a million words consisted of textbooks which were on an official reading list for students of science or technology at tertiary level. A vocabulary listing was collected for each chapter of the texts in the corpus. All closed-class vocabulary items were eliminated from consideration, and all remaining items were lemmatised. The collocational behaviour of the resulting set of lemmata was quantified by producing a listing of the frequencies of collocation of each lemma with every other lemma. The CLOC package[5] provided details of collocational frequencies within a span of up to a maximum of 12 words on either side of the node (the word originally searched for).

With this method, each lemma was characterised by the set of values comprising its frequencies of collocation with its collocates. For example, in one particular text the lemma *code* might collocate with *computer* three times, *encrypt* five times, *machine* zero times, *message* four times and *secret* ten times. The collocational behaviour of *code* in that text could then be represented by the following vector of values: 3, 5, 0, 4, 10. Different vectors for different genres may be a means of distinguishing between homographs. If, in another text, we get a collocational vector for *code* of 7, 0, 10, 0, 0, this would show that in the first case the field of discourse is cryptography while in the second it is the machine code used in computer science.

The collocational data for all the terms in the vocabulary forms a square matrix symmetric about the leading diagonal (which runs from top left to bottom right), where f12 is the frequency of collocation of terms 1 and 2. The principal diagonal represents the frequency of self-collocation, the number of times a lemma collocates with itself as in the examples *I had had enough* and *strong acids and strong bases*, where a word occurs more than once within the concordance window both as node and collocate.

The similarity matrix data was normalised by transforming the raw collocation frequencies such that (a) the value of 1 was assigned to each cell on the leading diagonal and (b) all other frequencies of collocation were scaled to fall within the range 0–1. Thus, no node was considered more similar to another node than it is to itself. The formula by which normalisation was achieved was $f_{ij}/(f_i + f_j - f_{ij})$ where f_{ij} is the number of times two terms collocate, f_i is the frequency of the first term and f_j is the frequency of the second term. For example, consider a collocation matrix of just three terms, such as the one shown in Table 3.14(a). Nodes 1 and 2 collocate with each other five times but none of the nodes collocates with itself. Node 1 collocates three times with node 3 while node 2 collocates four times with node 3. The values in Table 3.14(a) can be normalised to produce the values in Table 3.14(b). All the entries on the main diagonal, corresponding to self-collocation are set to 1. The non-normalised frequency of collocation for nodes 1 and 2 is five. Node 1 occurs a total of eight times, while node 2 occurs a total of nine times. The normalised collocational frequency, using the formula $f_{ij} /(f_i + f_j - f_{ij})$ is $5/(8 + 9 - 5) = 5/ 12 = 0.42$. Similarly, the normalised collocation frequency for node 1 and node 3 is $3/(8 + 7 - 3) = 0.25$, and that for nodes 2 and 3 is $4/(9 + 7 - 3) = 0.31$.

	Node 1	Node 2	Node 3
Node 1	0	5	3
Node 2	5	0	4
Node 3	3	4	0

Table 3.14(a) Non-normalised matrix of collocational frequencies

	Node 1	Node 2	Node 3
Node 1	1	0.42	0.25
Node 2	0.42	1	0.31
Node 3	0.25	0.31	1

Table 3.14(b) Normalised matrix of collocational frequencies

Due to storage considerations, the number of lemmata that were compared at once was limited by random sampling to produce a sample of about 60 items. The actual cluster analysis method employed in Phillips's study was Ward's method (Ward 1963). In order to evaluate this procedure, a pilot study was performed which successfully clustered terms known to be obvious collocates – for example, *quantum* with *mechanics* and *magnetic* with *field*. From the texts used for the main study, the following are extracts from the author's summary to the chapter entitled *Rotating Frames* (Phillips 1989, p. 53):

a) In problems involving a rotating body – particularly the earth – it is often convenient to use a rotating frame of reference.

b) These are the centrifugal force, directed outwards from the axis of rotation, and the velocity-dependent Coriolis force.

c) Rotating frames are also useful in any problem involving a magnetic field.

These quotations give some indication of the nature of the content of this chapter. The terms *centrifugal, velocity, field* and *axis* were omitted from the analysis by the random selection procedure, and the main lexical sets retrieved were as follows:

1. *force, Coriolis, order, neglect, term, third, acceleration, gravitational.*
2. *rotating, frame, reference.*
3. *angular, precess, constant, direction, swing, given, clearly.*
4. *magnetic, uniform.*
5. *earth, surface.*

This patterning of orthographic words reveals the lexical organisation of the text and is detectable on the scale of the whole text. This gives rise to the notion of the lexical macrostructure of texts.

5.3 Morphological analysis of chemical and medical terms

The words in a corpus can be annotated with semantic information (see McEnery and Wilson 1996) as a precursor to the clustering of terms according to their semantic similarity. One way in which semantic codes can be automatically assigned to the terms in a corpus, in particular one made up of journal articles in pharmacology, has been described by Oakes and Taylor (1994). Morphological analysis was performed on the organic chemicals subtree of the *Derwent Drug File* (DDF) *Thesaurus*, in which the names and structures of the chemical compounds are well correlated. The term morphology refers to the study of the make-up of words, where the smallest syntactic unit or meaningful word fragment is the morph. The set of morphs representing a given grammatical element is known as a morpheme. The meanings of many medical and chemical terms do seem to be directly derivable as the sum of the meanings of their constituent morphemes. Morphemes can be represented by a code number or a canonical lexical form.

These codes can be derived through morphosyntactic analysis, an extension of stemming rules, where the primitive concepts within a term are found by recognition of its constituent lexical morphs. The terms in a thesaurus are arranged in a hierarchy, and thus terms are also able to inherit concepts (represented by codes) from superordinate or parent terms. Leech (1975) states that a more specific term is said to be the hyponym of its more general term, and it contains all the features present in the parent term.

With stemming rules, the word fragments which are deleted or replaced do not necessarily correspond with standard grammatical prefixes and suffixes, and no record is kept of these word fragments as they are identified and manipulated. An extension to stemming rules is to restrict affix removal to those word fragments which correspond to word morphemes. By replacing each morpheme by a semantic code or codes, each corresponding to a predefined domain-specific primitive concept, we may perform morphosyntactic analysis on terms.

Morphological analysis is initially performed on each term in the organic chemicals subtree of the DDF database. This yields a vector of primitive codes for each concept, which is then augmented by including further semantic features determined by inheritance, whereby all child concepts in the thesaurus acquire all semantic features possessed by their respective parent concepts. User input terms may also be transformed into attribute vectors based on identification of their constituent morphemes, and then compared by Dice's similarity coefficient with the system's known terms, leading to a ranked output of related thesaurus terms.

A morpheme dictionary was produced, where each recognised morph in the organic chemical subtree was represented alongside its canonical lexical form and its primitive code or codes. Examples of morphemes assigned to each subgroup of the chemicals section of the DDF thesaurus were as follows: amino acids and peptides were assigned the morphemes *carboxyl* and *amino* and arenes and other benzenoids were assigned *aromatic* and *cyclic*.

The set of chemical primitives employed by the chemical terms help-system included numerics such as the Latin and Greek numerals up to 44, and *meth-*, *eth-*, *prop-* and *but-*, which refer to the length of a carbon chain, and were assigned codes 1 to 4 respectively. Other primitive codes were assigned to the categories given in a chemistry text book, groupings of compounds found in the DDF thesaurus, and concepts found in a chemical dictionary.

If the user inputs the term *aminocyclitol*, the identified components of the input term will be

1. the component *amino*, meaning *amino group* and given the code 57
2. the component *cycl*, meaning *cyclic* and given the code 50
3. the component *itol*, meaning *alcohol*, and given the code 93.

The stored version of *aminocyclitol* also obtains the above codes by morphological analysis, and in addition inherits the codes 66 and 69 which are the primitive codes for the parent term *monovalent, N-containing*. Dice's similarity coefficient is used to match the primitive codes of the input term to the code vectors of stored terms. If the top three matching terms are displayed, the output will be *aminocyclitol 0.75*; *peptide, cyclic 0.67*; *pyrrolidine 0.67*. In the case of *pyrrolidine*, the term inherits codes 50, 51, 5 and 69 as its parent term is *carbamates, ureas, 5-members, N-containing*, and its constituent morphemes are *pyrr-* (code 69), *-ol-* (code 93) and *-idine* (code 57). The same code derived from two different sources is counted just once in the term code vector.

Thus, morphological analysis was used in the development of a term retrieval system for the organic chemicals subtree. This help-system allows the user to input any chemical term, partial or complete, fully or partially recognised by the system, and the closest database known terms will be printed out in rank order of closeness to the input term. A related help system was also created, namely a translator of Greek and Latin medical terms. Primitive codes were not used in this case, but the everyday English canonical forms were printed out. For example, the input term *cardiopathy* would be translated by *heart* and *disease*.

5.4 Automatic thesaurus construction

One use of clustering terms according to their semantic similarity is in automatic thesaurus construction. Fully automatic thesaurus construction methods are based on the use of a set of document vectors, such as the one shown in Table 3.15, where each row corresponds to a particular document and column *j* of each document row shows whether term *j* has been assigned to each document. The similarity between pairs of columns is used to derive the similarity between term *k* and term *h*. If t_{ik} indicates the weight or value (such as the number of occurrences in the document) of term *k* in document *i* and if there are *n* documents in the collection, a typical term–term similarity measure is given by

$$S(\text{term } k, \text{term } h) = \sum_{i=1}^{n} t_{ik} t_i$$

	Term 1	Term 2	...	Term *t*
Document 1	d_{11}	d_{12}	...	d_{1t}
Document 2	d_{21}	d_{22}	...	d_{2t}
:	:	:		:
Document 3	d_{n1}	d_{n2}	...	d_{nt}

Table 3.15 Matrix of document vectors

A normalisation factor can be used to limit the term–term similarities to values between 0 and 1. A new matrix, shown in Table 3.16, can be created of all the term–term similarities calculated in this manner, in which the value in row *k* and column *h* is the similarity between term *k* and term *h*. Salton and McGill (1983) recommend that in order to form thesaurus classes, one should start with this term–term similarity matrix and perform single link clustering.

	Term 1	Term 2	...	Term *t*
Term 1	S(Term 1, Term 1)	S(Term 1, Term 2)	...	S(Term 1, Term *t*)
Term 2	S(Term 2, Term 1)	S(Term 2, Term 2)	...	S(Term 2, Term *t*)
:	:	:		:
Term *t*	S(Term *t*, Term 1)	S(Term *t*, Term 2)	...	S(Term *t*, Term *t*)

Table 3.16 Term–term similarity matrix

6 CLUSTERING OF DOCUMENTS ACCORDING TO SUBLANGUAGE USING THE PERPLEXITY MEASURE

A sublanguage is the semantically restricted subset of language used by a particular group of speakers, who might share a common interest such as birdwatching (Ashton 1996) or be employed in a common specialised occupation such as hospital surgeons (Bross, Shapiro and Anderson 1972). A sublanguage is characterised by such features as limited subject matter, 'deviant' rules of grammar, a high frequency of certain constructions and/or the use of special symbols (Lehrberger 1982).

Sekine (1994) used a newspaper (2147 articles of the *San José Mercury*) in his experiments on automatic sublanguage definition and identification. Each article in the newspaper was regarded as a unit of data, and a sublanguage would be formed by gathering similar units in terms of word appearance. This is similar to the text clustering technique in the field of information retrieval (Willett 1988). However, when creating clusters which are useful for inform-ation retrieval purposes, the linguistic features of those clusters are relatively unimportant. In contrast, Sekine's purpose was to find sublanguages useful for natural language processing systems, and so the linguistic features of the clusters were of importance.

In automatic sublanguage definition it is important to determine the number and size of clusters automatically, because otherwise one would need human intervention or the imposition of artificial thresholds. In order to achieve this, Sekine made use of the measure called perplexity, described in Chapter 2, Section 2.11. In order to calculate perplexity for a set of texts, the set must be treated as a single text. Sekine (1994, p. 125) states that

> if two texts have a large number of overlapping tokens, the perplexity of the combined text will be smaller than that of a text which has tokens chosen at random from the entire corpus. In short, if the perplexity for a text set is small in comparison to the perplexity for a random text, we may say that the set has sublanguage tendency.

Clusters were grown from one initial article by adding similar articles in order of their similarity to the initial article, using a similarity measure based on the logarithm of inverse document frequency. The estimated perplexity value after each addition of an article was compared with the perplexity of a sample of the same size of random text. Sekine plotted a graph, reproduced in Figure 3.4, of perplexity ratio against the number of tokens in the text. This ratio tends towards one as the number of tokens becomes high, because the cluster is becoming similar to the set of all articles combined. The perplexity ratio reaches a minimum value when a small number of similar articles are combined. The articles combining to produce this minimum value were considered to constitute a sublanguage. New articles could be assigned to sublanguage clusters derived in this way by assigning them to the cluster containing the most similar article using the log inverse document frequency ratio.

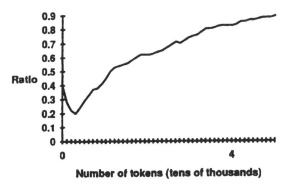

Figure 3.4 Sekine's data: perplexity ratio as a function of number of tokens

7 CLUSTERING OF DIALECTS IN GAELIC

As well as clustering terms and documents, clustering techniques may be employed in linguistics for the clustering of related dialects and languages. Kessler (1995) performed an experiment in which dialects of Gaelic were clustered. The original data for this study had been collected by Wagner (1958), who administered a questionnaire to native speakers of Gaelic in 86 locations in Ireland, the Isle of Man and seven locations in Scotland. The respondents had to provide the Gaelic word they would use for a variety of concepts, and these were transcribed into an alphabet based on the International Phonetic Alphabet. The first 51 concepts studied by Wagner were used in Kessler's experiment.

The first task in this cluster analysis was to compute the linguistic distance between each pair of locations. The measure selected was unweighted Levenshtein distance between the sets of phonetic transcriptions of concepts in each of the two locations. The unweighted Levenshtein distance is the cost of the least expensive set of insertions, deletions or substitutions that would be needed to transform one string into the other (Sankoff and Kruskal 1983). The simplest technique employed by Kessler, in which all operations cost one unit, was called **phone string comparison**.

This method was extended to produce a metric called **feature string comparison** which assigns a greater distance to substitutions involving greater phonetic distinctions. To account for all the distinctions in Wagner's original word lists, twelve distinct phonetic features must be taken into account: articulator, glottis, height, laterality, length, nasality, palatisation, place, rounding, strength, stricture and syllabicity. A phone was scored on an arbitrary ordinal scale according to how it corresponded with each of these features. For example, the values for place were glottal = 0, uvular = 0.1, postvelar = 0.2, velar = 0.3, prevelar = 0.4, palatal = 0.5, alveolar = 0.7, dental = 0.8 and labial = 1. The cost of substituting any two phones was taken to be the difference between the feature values, averaged across all 12 features. However, the distance matrix for the phone string comparison was found to be more closely correlated to Wagner's **isogloss** (contour map dividing regions in which different words for a concept are used) than the feature string comparison.

The feature string comparison distance matrix was used as the basis for bottom-up agglomerative clustering, which yields a binary tree of dialect locations. This type of clustering gave results which were well correlated with the isoglosses, and compared well with previously published accounts of dialectology in Gaelic.

Kessler states that one way of measuring how well a binary clustering technique works for dialect grouping is to compare for each site i its average dissimilarity from the other sites in the same dialect $a(i)$ with its average dissimilarity from the sites in the other dialect, $b(i)$. Kaufman and Rousseeuw's statistic $s(i)$ is defined to be $1 - a(i)/b(i)$ if $a(i)$ is less than $b(i)$, otherwise $b(i)/a(i) - 1$. The statistic thus ranges from 1 (site i fits perfectly in its assigned group) to -1 (site i would perfectly fit in the other group). This statistic allows one to visualise how well classified each site is, and also by finding the average value of this statistic across all sites, to derive an estimate of the quality of the clustering.

In conclusion, Kessler states that dialect groupings can be discovered objectively and automatically by cluster analysis. When agglomerative clustering based on the unweighted Levenshtein distance between phonetic strings is applied to Gaelic, plausible dialect boundaries are obtained, corresponding to national and provincial boundaries. This method is a great deal less tedious than deriving thousands of isoglosses. Similar studies have been performed by Schütz and Wenker (1966) on the dialects of Fiji and Évrard (1966) on the Bantu languages.

8 SUMMARY

Principal components analysis and factor analysis are included in this chapter on clustering because they cluster the variables used to describe a data item. These techniques summarise the information in a complete set of variables using fewer variables called principal components or factors, and hence reduce the dimensionality of the data without significant loss of information. For example, Horvath started with the relative occurrences of 25 vowel sounds to represent the speech of Sydney speakers, and these variables were reduced to four principal components. Once the factors or principal components are known, data items such as speakers or texts can be clustered according to their scores on the variables which go to make up the factors or principal components. Mapping or ordination methods can be employed to represent the relationship between more than two variables on a two-dimensional map.

A range of clustering methods were described with particular reference to document clustering. The two main methods are non-hierarchic and hierarchic clustering. Hierarchical clustering produces a set of clusters where smaller clusters of very similar documents are included within larger clusters of less similar documents, while with non-hierarchical clustering the resulting clusters are not included one within the other. For both types of clustering a similarity metric is required, either for calculating interdocument similarity or for calculating the similarity between a single document and a cluster of documents represented by a class exemplar. The four main types of similarity metrics are association, correlation, distance and probabilistic coefficients. The most common clustering methods for documents are hierarchical agglomerative methods, where the smallest clusters are created first and then merged to produce the larger encompassing clusters. In each case the starting point is a matrix of interdocument similarities. The methods (average linkage, complete linkage, nearest neighbour and Ward's method) vary in the choice of similarity metric and in the method of updating the similarity matrix each time a new cluster is formed.

Terms may be clustered according to lexical or semantic similarity. Lexical clustering of terms is performed by a range of approximate string matching techniques, which include term truncation (the similarity of two terms being measured by the number of initial characters they have in common), suffix and prefix deletion and substitution (stemming) rules, clustering of terms according to their constituent n-grams (sequences of adjacent characters), dynamic programming (the least number of insertions, deletions and substitutions required to transform one word into another) and the SPEEDCOP and Soundex coding systems. The applications of approximate string matching are in spelling checkers, automatic sentence alignment in parallel corpora and concordancing, and the identification of historical variants of words in a corpus. With regard to the clustering of terms according to semantic similarity, they can be clustered according to the constituents of concordance lines in which they occur, or according to their collocates in science texts. The fact that similar terms will occur in similar documents can be the basis for automatic thesaurus construction. Other types of clustering are the clustering of texts in a corpus according to the sublanguages they are written in, and the clustering of dialects and languages on the basis of common or related vocabulary.

The clustering techniques we have described are amongst the most effective available and their position within corpus linguistics is quite central. Increasing numbers of studies are being produced which use clustering techniques such as principal component analysis even though they may not identify them as such. Consequently, the techniques we have covered here will be returned to from time to time as we review other work in corpus linguistics. The close study of clustering techniques presented in this chapter should provide a firm footing for understanding their use in corpus linguistics.

9 EXERCISES

1. The similarity matrix for four Malayo-Polynesian languages has been estimated as follows (using the method of Chapter 5, Exercise 4):

	Javanese	Malay	Madurese
Sundanese	0.35	0.75	0.58
Javanese	–	0.35	0.82
Malay		–	0.45

Use these values to construct a dendrogram using the average linkage method.

2. Using the approximate string matching techniques of (a) truncation, (b) Dice's coefficient for bigrams, and (c) dynamic programming where insertion, deletion and substitution all count as one operation, calculate the similarity between the words *school* and *Schüle*. To convert this dynamic programming value into a coefficient of similarity, divide by the number of characters in the longer word.

3. For the word pair *linguistics* and *lists*, calculate (a) the longest common substring and (b) the longest common subsequence. Use these values to calculate Coggins's D_3 measure for this word pair.

4. Using the stemming rules of Paice (1977), reduce the following words to their lemmas: (a) *solution*, (b) *soluble*, (c) *solve*, (d) *solvable*. Which three form an equivalence class?

10 FURTHER READING

Numerical Taxonomy by Sneath and Sokal (1973) gives clear descriptions of many different clustering techniques, mainly applied to the field of biology. However, there is a brief but interesting section on the uses of clustering in linguistics. Sankoff and Kruskal (see Kruskal 1983) describe string matching techniques in various fields of science, including the analysis of human speech. Willett (1988) has published a comprehensive review of hierarchical document clustering techniques.

NOTES

1. See Morrison (1990) for a method involving matrix algebra.
2. Further details of the differences between PCA and FA in linguistics studies can be found in Alavi (1994).
3. See Paice (1977) for further details.
4. See McEnery and Wilson (1996) for a discussion of parallel corpora.
5. The CLOC package is described by Butler (1985b).

Concordancing, collocations and dictionaries

1 INTRODUCTION

Concordancing as such is of secondary interest for this book, being merely a means of gathering data that can be exploited, but concordancing as the process of gathering the data in its own right is a more central concern of the books by Barnbrook (1996) and Ooi (1998). The reader interested in knowing more about retrieval from corpora should refer to these.

A **concordance** is a list, arranged in an order specified by the user, such as the order of appearance, of the occurrences of items in a source text, where each occurrence is surrounded by an appropriate portion of its original context. Before the statistics used on concordances are introduced, the concordance itself will be described in Section 2 and Sections 2.1 to 2.4 will describe the typical output of concordances, decisions regarding the context or span of a concordance window, the preparation and annotation of texts for concordancing, and modes of sorting the output. Concordance packages requiring only very simple statistics, such as word counts – for example, COCOA, OCP, EYEBALL and WordCruncher – will be described in Sections 2.4.1 to 2.4.2.

Closely related to the concept of a concordance is that of the **collocation**. Collocations are groups of words which frequently appear in the same context, displayed, for example, around a keyword by a concordancer.[1] The extraction of collocations from corpora enables the creation of dictionaries for many purposes – guides for learners of a second language (Milton 1997) or technical glossaries (Daille 1995), for example. Collocations can be extracted using purely syntactic criteria, as described in Section 3.1, by observing regular syntactic patterns which are known to be typical of idiomatic collocations or technical terms. It will be shown that some parts of speech are more likely to form collocations than others, and that collocations are often enclosed by frameworks of characteristic function words. In Sections 3.2 to 3.2.14, a range of statistical measures used to identify collocations will be described in considerable detail. The themes covered will be:

- the retrieval of collocational word pairs
- the retrieval of technical **terms** from monolingual and bilingual corpora
- corpus alignment for the extraction of bilingual collocations
- extraction of collocations consisting of three or more words
- determination of the sense of ambiguous words by the set of collocates of that word
- extraction of collocations by a combination of syntactic and statistical information.

Further statistics useful in the production of dictionaries from corpora are measures such as Dunning's log likelihood measure (see Section 3.2.15) which shows if a word or phrase is overused or underused in a specialised corpus compared with a corpus of standard English, and **dispersion measures**, which show how evenly a word is distributed in the corpus (see Section 3.2.16). A multivariate technique which shows the different collocational behaviour of a word in different corpora is Hayashi's quantification, described in Section 3.2.17.

Having described the statistics for identifying collocations, some concordance packages which make use of these statistics will be described in Sections 4.1 to 4.4, in particular WordSmith and CobuildDirect.

2 THE CONCORDANCE: INTRODUCTION

A concordance lists, in a suitable order, the occurrences of items in a source text, where each occurrence is surrounded by an appropriate portion of its original context. Concordances are the oldest and most common application of computers in the humanities (Oakman 1980). The earliest published concordance produced photographically from computer printout was a concordance to the poems of Matthew Arnold by Parrish (1959).

The first step in the generation of a concordance is to produce a **word index** from the source text. Hockey (1980) defines a word index as simply an alphabetical list of the words in a text usually with the number of times each individual word occurs and with some indication of where the word occurs in the text. Such a word index is known in the field of information retrieval as an **inverted file** (Salton and McGill 1983), an example of which is shown in part in Table 4.1. The entry for 'anaphoric' shows that the word occurs twice in the source text (the value before the colon), and these two occurrences are found in sections 52 and 53 of the source text, where a section may be a line or a paragraph.

algorithm	3:	125, 134, 144
anaphoric	2:	2, 52, 53
annotation	3:	24, 57, 119
artificial	3:	12, 16, 119

Table 4.1 Part of an inverted file

When each occurrence of each word (called the keyword, headword or sort key) is presented with the words that surround it, the word index becomes a concordance. This concordance is often called a KWIC, *Key Words in Context*, concordance. Part of a KWIC concordance showing the keyword *clowns* centrally aligned is shown in Table 4.2. Sometimes the entire concordance for a book or corpus is published, but in other literary studies only the section of the concordance for one or selected number of keywords is printed out.

in sterling as in punts, and bands and	\<clowns\>	to provide extra entertainment. Racing
lement, brilliant trick biker, ridiculous	\<clowns\>	, sublime running gags, over-hyped, u
arrived in military trucks, dressed as	\<clowns\>	. Authorities have themselves acknowl
circus, the Circo Price, about whose	\<clowns\>	I wrote a poem in The Prodigal Son. I
er and such well-known and beloved	\<clowns\>	as Grock, Ramper and Charlie Rivel a
spkr\> Jeeves: \</spkr\> Some of those	\<clowns\>	should get back to Billy Smart's ! \</s
s lost but now I 'm found Send in the	\<clowns\>	Sincerely yours A song like I'm singin
be bribed with the prospect of circus	\<clowns\>	and Charlie Chaplin films. They'd slip
laying himself for once in a crowd of	\<clowns\>	, and though he could only be real in a
seemed not the amusing and lovable	\<clowns\>	they were meant to be but somehow

Table 4.2 Part of a concordance produced from the British National Corpus

One way to view the word index is as a concordance with no printed context (Hays 1967). With it, the user must refer back to the source text to check the usage of a word in its original context. The advantage of the simple word index over the concordance is that it is not necessary for the machine to estimate the required amount of context beforehand, since this is effectively selected by the index user when the source text is consulted. On the other hand, there are many ways to elaborate on the concept of the concordance to make it more useful. For example, the sort key need not be a single word, and there are various ways in which features in the environment of a sort key can control the ordering of occurrences of the sort key in the concordance. In order to limit the output of a concordance, it may be best to place the most common words in an exclusion list or **stoplist**, so that no concordance output is produced for those words. However, as high-frequency words are often of most interest to those studying linguistic usage, another class of words which are often stoplisted are the **hapax legomena**, the lexical units which occur just once each.

Concordances have a wide range of applications. Rudall and Corns (1987) list the following examples of possible functions of a concordance in literary stylistics:

1. As a memory aid. A partly-remembered quotation can be located in the original text by first looking up the remembered part in a concordance.
2. A concordance can be used as an index, since the original location of each concordance line (such a the scene or line) is normally given.

3. The concordance facilitates approaches of linguistic stylistics to the study of both lexis and syntax. For example one can pick out and classify compound words, neologisms, archaic and slang words, or words drawn from the native vocabulary of English. Texts can be characterised as containing many such words. Concordances can assist the user in identifying homographs of a word, which is not easy by purely automatic methods. Sorting words by their endings rather than by their initial characters enables one to classify words by part of speech and to count the occurrence of each part of speech.

4. Concordances enable thematic analysis, or the identification of clusters of words which distinguish the linguistic universe of the text, such as words on the theme of light. Similarly, one can examine whether two different texts or two different parts of a single text differ, such as the parts spoken by two different characters in a play. The original text of a play can be annotated so that the concordance can distinguish different speakers.

Concordances are also valuable tools for observing collocations or textual items in the sort key environment which consistently appear in conjunction with a given sort key. Beyond these uses of a concordance, however, lies the possibility of statistical processing of concordance data. We will return to this point in Section 3.2.

Before we consider context in a concordance any further, however, we must briefly consider a relatively new type of computer concordance, a bilingual concordance. A bilingual concordance can be produced by aligning parallel text in two different languages so that a line from one text is followed immediately by the corresponding line from the other text. A pair of lines can then be specified as the context of any word (in either language) appearing within them. One application of a bilingual corpus is to examine which instances of a word in one language are translated in one way, and which in others, as performed by Gale, Church and Yarowsky (1992) and described in Section 3.2.12.

2.1 Determining the context of a concordance
De Tollenaere (1973) discusses the problem of the context in computer-aided lexicography, where dictionary words are examined in context to show the different senses in which they may be used. To avoid having to refer back to the original text every time in order to determine the meaning of a word, the dictionary-maker must be able to view sufficient context in the concordance for the task at hand. A context printout which is too long may be abridged by the lexicographer in a few seconds, but a too short context will take a long time to complete. De Tollenaere suggests that a computer-generated·context might, for example, consist of four lines before the headword and two after. This reflects the fact that the significant or 'defining' context usually precedes the

keyword, and that the two lines following the keyword in most cases will allow the context to be syntactically complete. Shorter, mechanically determined contexts rarely correspond with syntactic units.

There are four basic strategies for determining the context of a concordance, or the number of lexical items surrounding each keyword. Firstly, one can stipulate a specified number of characters around each keyword, but this has the disadvantage of splitting those words which occur at the beginning or end of a concordance line. Secondly, each keyword can be given in the context of the line in which it occurs. This method is particularly good for presenting concordances of verse. Thirdly, the context may be the words around the keyword which are bracketed by punctuation. In this case the output may be unwieldy, and it may be necessary to set a maximum number of words for the context. A fourth, labour-intensive, possibility is the manual post-editing of a larger context produced by the machine. The option of pre-editing texts to determine contexts is also too labour-intensive for large texts (Rudall and Corns 1987). In practice, commercial concordance packages tend to use the first option.

The reality is that refinements for all of these strategies are possible. For example, consider the third option. De Tollenaere suggests the use of a hierarchy in punctuation, the order of precedence being full stop, exclamation mark or question mark, followed by semicolon then colon and finally comma. The concordance program should search between 180 to 120 characters before the headword for a punctuation mark of the highest precedence, to begin the context for the concordance output at that point. If none of these punctuation marks is found, the program should scan again for a semicolon, and so on. The same procedure could be used to close the context. Such proposed refinements are legion. In spite of this, the basic strategy most widely used is to simply specify 'n words on either side'.

2.2 Text selection and preparation for concordancing

Rudall and Corns (1987) discuss ways in which the corpus can be pre-prepared to prevent orthographically distinct instances of the same word, such as *be* and *was*, being listed separately. In a small concordance for private use it might be possible to normalise the spelling in the corpus, but for a publicly available concordance one must either use a widely current and respected version of the text or go back to the originals of the text. This will involve more work, but allows one, for example, to bring together variant readings of a text into a single corpus. The identification of variant spellings can be facilitated by generating a token list from the corpus, containing one entry for each word which occurs in the corpus irrespective of its frequency.

Another preprocessing task is to insert into the text file the references which are to be included in the concordance to identify the location of each keyword in the corpus – the date or the genre of the text, or speaker in a spoken corpus, for example. At the start of each play or other textual unit a marker must be inserted, distinguished from the raw text (perhaps by an SGML entity), so it is not processed itself when the concordance is produced.

Preprocessing of the corpus may also be necessary to cater for lemmatisation and homograph distinction. A simple concordancer for an unannotated corpus will separate the various inflected forms of the same word. Sometimes inflected forms will be sorted close together as in the case of *dog* and *dogs*, but sometimes the listings will be further apart as for *man* and *men*, *go* and *went*. Some examples of the same word spelt differently will also be sorted far apart, such as *enquire* and *inquire*. However, homographs will always be sorted together. Five strategies suggest themselves to overcome these problems:

1. Pre-editing of the corpus, such as by inserting markers to distinguish homographs as in *lead1* and *lead2*. This may be difficult for some poetry, where double meanings of a word are sometimes used.
2. Enhancing the concordance program by instructing the machine to sort together groups of inflected or orthographically distinct words. Automatic lemmatisation may bring the various senses of a word together, except where prefixes and infixes are used in declensions.
3. Post-editing of the concordance. Items listed apart can be moved together if desired, or entries of homographs sorted together can be separated according to their contexts.
4. Cross-referencing and annotation. For each headword, a note should be given to the user if there are known to be alternative forms and closely-connected words which should also be looked up.
5. Approximate string matching to group spelling variants (Robertson and Willett 1992), described more fully in Chapter 5.

There are various modes in which corpora can be annotated for concordancing. Using dictionary look-up, a thesaurus code can be assigned to each word, enabling words with similar meaning to be sorted close together. Using morphological information, words can be divided into their roots and affixes. One can then study the frequency distribution of various affixes, or conversely sort all forms of the same baseword together. Syntactic information can be appended to the corpus by automatic parsing, or the annotation of dependency information, where the keyword can be sorted according to what it governs (its dependent) or what governs it. In other words, the context used in sorting can be fixed syntactically, rather than by spatial adjacency. For a comprehensive overview of corpus annotation see McEnery and Wilson (1996) or Garside, Leech and McEnery (1997).

It would also be desirable for a concordancer to decide if there are sections of the corpus which should be treated separately, such as quotations from another author, stage directions or foreign words, since these may distort vocabulary counts. Such sections can be marked by special characters in the text such as square brackets, and one can create a separate concordance just for these words. SGML annotated corpora and retrieval software make this feasible.

2.3 Modes of sorting the output from a concordance

Hockey (1980) lists the orderings in which the occurrences of a word might be listed by a concordance. The most usual is the order in which they occur in the text. By using the alphabetical order of what comes to the right or left of the keyword, the instances of a particular phrase can be brought together. It is also possible to reverse the sort keys, so that they become sorted in alphabetical order of their endings. After sorting, the sort key can be turned around again before printing, producing what is called a 'reverse concordance', useful in the study of morphology, rhyme schemes and syntax. The computer can sort words in any alphabet order. For example, the Hua Xia concordancer of Oakes and Xu (1994) could sort Chinese characters by alphabetical order of their romanised (Pinyin) equivalents.

A more elaborate and effective mode of sorting according to the left and right components is described by Altenberg and Eeg-Olofsson (1990) as **zig-zag** sorting. Their aim was to identify recurrent word combinations in the London–Lund corpus of spoken English. Simply finding the frequency of all word pairs, word triples and so on would not be adequate, since recurrent word combinations vary in length and frequently overlap. Their chosen procedure was to sort their concordance in zig-zag order: first according to the keyword, then according to the first word to the right, then the first word to the left, the second to the right, and so on. This resulting layout was suitable for manual inspection, and a short sample is shown in Table 4.3. Using zig-zag order also makes it easy for computer programs to retrieve all recurrent combinations containing a certain keyword, where the instances of recurrent word combinations are indexed on their central words.

past five probably	a	bit after that
have been pruned	a	bit more
fact there's been quite	a	bit of painting this
reckon we spent quite	a	bit of time just wandering
in fact there's quite	a	bit of reflection
which takes up quite	a	bit of time
is really	a	bit excessive

Table 4.3 Sample of zig-zag sorted concordance of recurrent word combinations in the London–Lund corpus (keyword = *a*).

To refine the retrieval process, Altenberg and Eeg-Oloffson eliminated all included instances: all word combinations included in larger ones (such as *a bit* being included in *quite a bit of* were automatically weeded out. The next step was the manual elimination of phraseologically irrelevant examples (those which were not linguistically interesting), such as *of the*, *it a*, or *the whole he*. The remaining word combinations were then analysed in terms of grammatical type using automatic tagging and parsing programs, and in terms of function, categorising combinations as characteristic of conversation types such as face-to-face or telephone conversation.

With a hyphen, various output schemes are possible. Treating it as a character will bring all instances of *lady-like* immediately after a block of instances of *ladylike*. If it were ignored completely, both forms of the word would be mixed in the same block. If it is regarded as a space-like delimiter or word separator, the two halves would be considered as separate and distort the word counts for *lady* and *like* (Hockey 1980). Producing a definitive response to this problem is not really possible as the acceptability of each option is somewhat user-specific.

Finally, let us consider sorting on word ending. Wisbey (1971) produced a concordance of rhymes in Old German poetry. Each word occurring at the end of a line was cross-referenced by listing its total number of occurrences, the location of each occurrence, and the word it rhymed with at that location. A line of such an index might read *mir (2) 228 dir 400 ir*, showing that *mir* occurs twice, once at the end of line 228 where it rhymes with *dir* and once at the end of line 400 where it rhymes with *ir*. Since Wisbey was compiling an index of rhymes, the entries were sorted in reverse alphabetical order.

2.4 Simple concordance packages

Now that the basics of concordancing have been discussed, in the following sections (2.4.1 to 2.4.4) four different concordance packages will be described: COCOA, OCP, EYEBALL and WordCruncher. They are grouped together, since although they offer a range of facilities for collating and sorting the output, they are simple in the sense that they do not use any advanced statistical measures. Their statistical output is restricted to such measures as the frequency of each word in the corpus, frequency tables showing how many words appear once, twice and so on and a type-token ratio. Later in this chapter, after more sophisticated measures of lexical statistics have been introduced, four concordance packages which use them (Wordsmith, CobuildDirect, Lexa and Hua Xia) will be described.

2.4.1 COCOA

COCOA (COunt and COncordance on Atlas) was originally developed at the Atlas laboratory. It allows alphabetisation of the context on the word to the left or right of the keyword (Oakman 1980) and permits the creation of a user-specific alphabetical order. For example, in Spanish word lists the letter pairs *ch* and *ll* are normally sorted after *c* and *l* respectively. COCOA enables such a sorting order to be specified, so *llano* comes after *luz* rather than before it as would happen naturally with an English alphasort. COCOA can also produce reverse alphabetical lists, where a reverse index makes it possible to study patterns of rhymes or word endings in inflectional languages. In highly inflected languages, morphological endings come together in reverse indexes. For example, Wisbey's concordances to Middle High German texts (Wisbey 1971) enabled the historical study of both stems and endings in the development of the German language. COCOA permits the selective concordancing of words that fall within a certain frequency range or begin or end with particular prefixes and suffixes, such as restricting the output to only those words ending in -*ing* that occur at least 10 times in the corpus. Hyphenated words like *a-dreaming* can be automatically sorted to appear in two places in the text, under both *a-dreaming* and *dreaming*. COCOA also allows the user to specify searches for the co-occurrence of two specific words within a set number of words, retrieving, for example, all sentences where *tropic* and *cancer* occur within two words of each other.

In terms of vocabulary statistics, COCOA is able to generate frequency tables which show how many words occur once, twice, three times and so on. The type-token ratio, which is the ratio of different vocabulary items to the total number of words, is available. This is useful, as it is a measure which can be used to differentiate genres and writers (Rudall and Corns 1987). The reference format employed by COCOA is to enclose the reference in angle brackets, and consists of a one-letter category identifier, where, for example, <T Catch 22> indicates the title of the novel, and <S3> indicates the third stanza of a poem (Butler 1985b). COCOA was a forerunner of the Oxford Concordance Program, described in the following section.

2.4.2 The Oxford Concordance Program

With the Oxford Concordance Program (OCP),[2] the sorting of keywords can be alphabetical, by ascending or descending word frequency, or according to word length (Hofland 1991). A list of prefixes and suffixes can be specified, which will then be ignored in the sorting process. The concordance can be sorted by right-hand or left-hand context. The required keywords can be stored beforehand in a list. This list may contain patterns where * stands for several or no characters and @ for exactly one character. For example, *cut** stands for *cut*, *cuts* or *cutting*, and *c@t* stands for *cat*, *cot*, or *cut*. Concordances can cope with word combinations (including patterns) or pairs of words with up to a specified number of undefined words in between. The OCP allows the specification of such factors as requesting that the letter *e* with different accents should be sorted together. The output of the OCP can be specified as a word list, an index or a concordance, along with simple vocabulary statistics (word frequencies, number of words in each frequency band, and type-token ratio) if required.

2.4.3 EYEBALL

In the EYEBALL concordance program each word is isolated and stored in an array which receives these additional pieces of information (Ross 1973):

- its location in the text
- the word length in syllables
- its syntactic category (such as noun, verb, determiner, or pronoun)
- its syntactic function (such as subject, predicate, complement or adjunct)
- a phrase–clause code which shows if the word ends a syntactic unit
- the word itself
- punctuation, if any, which follows the word.

Any of these pieces of information can be used as the sort key.

2.4.4 WordCruncher

WordCruncher[3] consists of an index creator called IndexETC and a browser called ViewETC (Hofland 1991). The user supplies a text corpus, which may be in English, French, German or Spanish, from which the indexer creates an inverted file consisting of each word in the corpus, the number of times it occurs and the location of each occurrence. Locations of words in the original corpus are described using a three-level system specified by the user, where, for example, a location might be specified by book (level 1), chapter (level 2) and sentence (level 3). The user can supply a stoplist of words which are not to be indexed. The sorting order of the various characters in the text may be specified, where, for example, upper- and lower-case characters may be specified as equivalent, or other characters such as hyphens are to be ignored in the sorting.

Once a text has been indexed, its lexical contents may be examined using the browser. After a text has been chosen, the main menu offers the options of looking up words or word combinations selected from a thesaurus or an alphabetical list of all distinct words occurring in the text. One can also view a frequency distribution showing how the usage of the chosen word varies according to the level 1 (broadest category) references, such as book or text genre. Alternatively, one can look up specific text references by specifying for example book, chapter and verse, or elect to make a KWIC concordance for selected parts of the text. It is possible to retrieve sequences of words, possibly with undefined words in between. One can specify that one word must occur before or after the other word, or either. Searches may be made in a tagged corpus, where, for example, one may retrieve all nouns (tagged using the syntax *word_nn*[4]) which occur within 30 characters of a form of the verb *to have* such as *has_hvz* or *having_hvg*. The ViewETC browser may be conveniently accessed from WordPerfect.

3 COLLOCATIONS: INTRODUCTION

According to Kjellmer (1990), words in natural language tend to occur in clusters. **Collocations** are such word groups which frequently appear in the same context. Smadja (1991) writes that neither syntax nor semantics can justify the use of these word combinations, and refers to them as 'idiosyncratic collocations'. These account for a large part of English word combinations. 'Idiomatic collocations' are phrases which cannot be translated on a word-for-word basis. Smadja, McKeown and Hatzivassiloglou (1996) write that published translations of idiomatic collocations are not generally available, even though a knowledge of collocations is very important in the acquisition of a second language. One reason for this is that collocations tend to be specific to a domain sublanguage, and thus the collocations used in a sublanguage often have different translations to those in general usage. The concept of collocations is closely related to that of concordance output, since the idea of two words occurring in a common context is similar to that of two words occurring in the same concordance window.

Altenberg and Eeg-Olofsson (1990) state that collocational information is relevant to foreign language teaching, since many student errors are best explained collocationally, and a knowledge of collocational behaviour means that new lexical items can be first introduced to the student in their habitual environments. Dictionaries of collocations can be created from monolingual corpora, although in some studies (such as Gaussier and Langé 1994) this process is facilitated by the use of bilingual corpora. According to Haskel (1971), words are attracted to each other to form collocations if they are of the same derivation, they express opposite meanings, or they commence with the same consonant clusters. Collocations enable stylistic variety, since the creative author can bring together disparate words, forming unorthodox collocations.

Smadja (1992) lists the following applications of programs for identifying collocations:

- Finding multiple word combinations in texts for indexing purposes in information retrieval, since these provide richer indexing units than single terms. Another way a knowledge of collocations can increase the precision of information retrieval systems is that a collocation gives the context in which a given word was used, which will help retrieve documents containing an ambiguous or domain-specific word which use that word in the desired sense.
- For automatic language generation, knowledge of collocations must be pre-encoded in a lexicon. Lists of collocations are difficult to generate because they are domain-dependent and idiomatic and come in a large variety of forms.
- Multilingual lexicography. A word-by-word translation of *semer le désarroi* (to cause havoc) would yield the incorrect sequence *to sow disarray*. This shows the need for bilingual collocational dictionaries, against which translations can be checked.
- Knowledge of collocations will also improve text categorisation systems.

Gledhill (1996) explored the link between collocations and technical terms. He found that in a corpus of cancer research articles the term 'management' only occurs in phrases such as 'patients received active management'. Since this can only be interpreted alongside more typical expressions such as 'patients received drug X' (where X is a treatment-related drug) it can be assumed that 'management' is a technical term for a course of drugs. Gledhill makes a distinction between technical terms and collocations which are phrases in everyday use, but authors such as Daille (1994) use techniques for finding collocations specifically to extract technical terms.

Collocations also have a role to play in translation. Nagao (1984) suggests that a human translator will first decompose a source sentence into phrases or fragments, then translate each fragment by analogy with other examples, then finally combine the target language fragments into a single sentence. It is possible that collocations correspond with these fragments which act as translation units. So what are we observing when we see a collocation? Wolff (1991) suggests that in order to maximise the efficiency of storage of lexical items in the brain, certain words are stored as patterns of more than one word, these word patterns appearing in language in a variety of contexts. Collocations might therefore correspond with such word patterns (Kita et al. 1994).

With this in mind, let us consider a definition of collocations. Kjellmer (1984) uses the following definition of collocations: collocations are both lexically determined and grammatically restricted sequences of words. *Lexically determined* means that in order to be considered as a collocation, a word sequence should recur a certain number of times in the corpus. *Grammatically restricted* means that the sequence should also be grammatically well formed according to Kjellmer's criteria. Thus, of the three sequences *try to, hall to* and *green ideas*, only *try to* is counted as a collocation, since *green ideas* occurs only once in Kjellmer's corpus and *hall to* is not a grammatically well-formed sequence. The distinctiveness of collocations is a matter of degree rather than an all-or-nothing feature. Kjellmer suggests that the following factors could be used to indicate the degree of collocational distinctiveness:

- Absolute frequency of occurrence, where the more frequent the collocation, the more distinctive. This criterion has been employed by many authors.
- Relative frequency of occurrence, being the ratio of the observed frequency of a collocation to the frequency expected if the words occurred together only by chance. This ratio is expressed by the mutual information formula (used, for example, by Church and Hanks 1990).
- Length of sequence, where, for example, the collocation *figured prominently in* seems more distinctive than simply *figured in*. Sequence length is incorporated into the 'cost criterion' formula of Kita et al. (1994).
- Distribution of the sequence over texts or text categories. This criterion may be evaluated using the measures of diversity, used by Daille (1994, 1995) and dispersion, used by Lyne (1985, 1986).

In the following section we will examine Kjellmer's assertion that a collocation is a grammatically restricted sequence, and in Section 3.2, measures of collocational distinctiveness will be described, including those which examine Kjellmer's criteria.

3.1 Syntactic criteria for collocability

In connection with collocation studies, Kjellmer (1990) asked whether different words or word types differ in their tendency to cluster. In the Gothenburg corpus of collocations, drawn from the Brown corpus (which will be referred to as the non-collocational corpus), collocations are defined as 'recurring grammatically well-formed sequences'. Using a grammatically-tagged version of both the corpus of collocations and a non-collocational corpus, he was able to establish which word classes are 'collocational' and which are not. In Table 4.4 the percentage of words with a given part-of-speech tag used in collocations and in non-collocations are given. These results show that the parts of speech AT, IN, NN and VB are collocational, while JJ, NP and RB are non-collocational.

Tag	Collocational corpus	Non-collocational corpus
AT article	17	0
IN preposition	16	1
JJ adjective	5	28
NN singular or mass noun	21	6
NP singular proper noun	4	21
RB adverb	1	21
VB verb, base form	3	0

Table 4.4 Tag percentages in collocational/non-collocational corpora

The ratio ((*occurrences in collocations* x *100*) : (*occurrences in the Brown corpus*)) will give an indication of the collocability of each word as a percentage. This 'collocational ratio' was computed for all the words in the Brown corpus with a frequency of four or more. Of words which could be unambiguously tagged, the 500 'most collocational' and the 500 'least collocational' were found and sorted according to tag, giving results similar to those shown in Table 4.4.

Common grammatical words combine with each other in various ways. Renouf and Sinclair (1991) discuss collocational structures or 'frameworks', consisting of discontinuous pairings of common grammatical words which 'enclose' characteristic words or groupings of words. They studied frameworks of pairs of high-frequency grammatical words with one intervening word or 'collocate'. Examples of frequent collocational types found for the framework *a + ? + of* in the written corpus were *a lot of, a kind of* and *a number of.* The variable collocates were mostly nouns. The frameworks studied, the total number of occurrences of each framework (tokens), the number of different intervening words found (types) and the type-token ratio for written and spoken data obtained from the Birmingham Collection of English Text are given in Table 4.5. The type-token ratio indicates how selective each framework is in its choice of possible collocates, lower values indicating greater selectivity.

Framework	Spoken corpus			Written corpus		
	Tokens	Types	Ratio	Tokens	Types	Ratio
a + ? + of	3830	585	6.0:1	25,416	2848	8.9:1
an + ? + of	208	94	2.2:1	2362	479	4.9:1
be + ? + to	790	216	3.6:1	5457	871	6.3:1
too + ? + to	59	36	1.6:1	1122	387	3.0:1
for + ? + of	127	56	2.3:1	1230	332	3.7:1
many + ? + of	63	36	1.8:1	402	159	2.5:1

Table 4.5 Frequency of occurrence of collocational frameworks

3.2 Measures of collocation strength

Given that we can spot collocational candidates, how might we rate their relative importance? The answer is to measure collocational strength. In the following sections we will examine a variety of measures of the strength of the collocation between two or more words, since these measures can identify such phrases as idiomatic collocations and technical terms in a corpus, which can then be collated into dictionaries. For the identification of significant two-word units in monolingual corpora we will examine Berry-Rogghe's (1973) z score and **R score** (Section 3.2.1), and the **C score** of Geffroy et al. (1973) which depends on both the frequency of a word pair and the proximity with which they occur in the text (Section 3.2.2). In Section 3.2.3 we will examine the use of combination theory in determining the significance of a collocation. In Sections 3.2.5 to 3.2.6, the work of Daille (1994) and Gaussier and Langé (1994) at c2v and IBM in Paris will be described, in which technical terms are extracted firstly from a monolingual corpus alone and then by making use of syntactic pattern affinities in a bilingual corpus. When bilingual corpora are used for term extraction, they must generally be aligned beforehand. After a description of an automatic alignment algorithm in Section 3.2.7 will follow a description of the use of bilingual corpora on term extraction on the CRATER project (McEnery and Oakes 1996). The use of collocation strength measures for the identification of phrases consisting of three or more words will be covered, for which the Cost Criterion, Factor Analysis and Luk's (1994) use of mutual information for the segmentation of Chinese words will be described (Sections 3.2.9 to 3.2.11). The XTRACT program, described in Section 3.2.12, uses a mixture of syntactic knowledge and statistics to extract collocations of two words from a monolingual corpus. Smadja, McKeown and Hatzivassiloglou's *Champollion* program (1996) then uses a bilingual version of the same corpus to assemble these word pairs into longer collocational phrases. The measure of collocational strength used for this purpose is Dice's similarity coefficient. In Section 3.2.15, statistics for determining whether a word appears more frequently in a genre-specific corpus than in a general corpus will be described. Finally, in Section 3.2.16, we will encounter dispersion measures, which show how evenly a word is distributed throughout the corpus, since Kjellmer (1984) states that idiomatic collocations tend to be spread throughout the text. With this done, we can move to an examination of the use of some of these measures in available concordance packages.

3.2.1 Berry-Rogghe's z-score calculation

Berry-Rogghe (1973) aimed to compile for each word in the corpus a list of its significant collocates. The phrase **significant collocation** can be defined in statistical terms as the probability of one lexical item (the node) co-occurring with another word or phrase within a specified linear distance or span being greater than might be expected from pure chance. In order to quantify this, the following data must be defined:

Z: the total number of words in the text.
A: a given node occurring in the text F_n times.
B: a collocate of A occurring in the text F_c times.
K: number of co-occurrences of B and A
S: span size; that is, the number of items on either side of the node considered as its environment.

The probability of B co-occurring K times with A, if B were randomly distributed in the text, must first be computed. Next, the difference between the expected number of co-occurrences and the observed number of co-occurrences must be evaluated. The probability of B occurring at any place where A does not occur is expressed by:

$$p = F_c / (Z - F_n)$$

The expected number of co-occurrences is given by:

$$E = pF_n S$$

The problem is to decide whether the difference between the observed and expected frequencies is statistically significant. This can be done by means of computation of the z score using the formula

$$z = (K - E) / \sqrt{Eq}$$

where $q = 1 - p$. This formula yielded a gradation among collocates with high scores assigned for collocations intuitively felt to be strong. The problem with this formula is that it does not allow a word to collocate with itself. Berry-Rogghe used the z-score formula to study collocations of *house* in a corpus made up of *A Christmas Carol* by Charles Dickens, *Each in his own Wilderness* by Doris Lessing, and *Everything in the Garden* by Giles Cooper. In some cases negative z scores were computed, showing that in some sense two words can **repel** each other. The sum of all the z scores for a given word should be zero. For a collocation to be statistically significant at the 1 per cent level, the z score should be at least 2.576. Berry-Rogghe repeated the experiment for different spans in the range 3 to 6 to find the optimal span. Increasing the span size resulted in introducing desirable collocates from *A Christmas Carol* but undesirable ones from the two plays.

When the span was set to three, the significant collocates for *house* were as follows, with the highest scoring given first: *sold, commons, decorate, this, empty, buying, painting, opposite, loves, outside, lived, family, remember, full, my, into, the, has.* In each case the z score was significant. When the span was set to six, the significant collocates were: *sold, commons, decorate, fronts, cracks, this, empty, buying, painting, opposite, loves, entered, black, near, outside, remember, lived, rooms, God, stop, garden, flat, every, big, my, into, family, Bernard, whole.* Thus, increasing the span removes the undesired collocates *the* and *has* which have no particular relation with *house*, and picks up several desirable collocates of *house* such as *rooms, God, garden* or *flat*. The only undesirable collocate to be picked up by increasing the span was *Bernard*. The majority of collocates occur within sentence boundaries, so a text with longer sentences (the average sentence length for *A Christmas Carol* was 14 words while for the plays it was only 6.7) has the greater optimal span. However, important collocates can be missed due to anaphoric reference. Overall, Berry-Rogghe found that the optimal span was four, using a variety of words, except for adjectives where the optimal span was just two. The optimal span could effectively be increased from four to six or seven items, by employing an exclusion list of words such as *and, but, it, its, nor, or, that, what, whether, which, who,* and all forms of the auxiliaries *be, do* and *have*. These words would be ignored in the calculation of all frequencies in the z-score formula and in determining the length of the span.

In 1974 Berry-Rogghe again used the z-score formula, this time for the automatic identification of phrasal verbs. Phrasal verbs were defined as idiomatic occurrences of a verb followed by a particle, as in *look after* or *give in*. Major syntactic criteria for the identification of phrasal verbs are as follows:

- Typically phrasal verbs are replaceable by a single item which is synonymous to it, as in *come up* = *mount, come in* = *enter,* or *come out* = *leave,* as opposed to non-idiomatic occurrences of verb + particle such as *look at* or *give to*.
- A phrasal verb can normally undergo passive transformation. For example, the idiomatic *They arrived at a decision* can be replaced by *A*

decision was arrived at, but the non-idiomatic *They arrived at the station* cannot be replaced by ★*The station was arrived at.*

- The phrasal verb cannot normally be inverted: for example the idiomatic *It stood out* cannot be replaced by ★*Out it stood*, but the non-idiomatic *He came out* can be replaced by *Out he came.*

All three criteria have exceptions, since idiomaticity is not absolute but rather a matter of degree.

Rather than obtaining collocational sets of every verb in the text, it was more practical to start by examining the particles which are limited, and to examine their left-order collocates. The head word *in* was chosen as it was the most frequent particle with 2304 recorded occurrences. Using COCOA, Berry-Rogghe produced a concordance of the keyword *in* sorted to the left, so, for example, *believe in* comes before *give in*. The underlying corpus was of 202,000 words, consisting of works by Doris Lessing, D. H. Lawrence and Henry Fielding. All left-hand side collocates of *in* whose *z* score was greater than three were found, which, in order of greatest *z* score first, were as follows: *interested, versed, lived, believe, found, live, ride, living, dropped, appeared, travelled, die, sat, died, interest, life, rode, stood, walk, find, house, arrived, came*. This list, which we will refer to as list 1, shows what we might intuitively expect, namely that words such as *interested* and *believe* are more closely associated with *in* than such words as *walk* or *sit*. However, this list of collocates ordered by decreasing *z* scores alone is not quite sufficient to quantify the degree of idiomaticity. For example *lived in* should be less idiomatic than *believe in*, though the list suggests otherwise.

An expression is said to be idiomatic when the meaning of the whole differs from that of the separate parts. This could be observed if the whole expression attracts a sufficiently different set of collocates to the collocates of each of the parts of the expression. Berry-Rogghe gives the example that *hot dog* collocates with *eat, mustard* and *stalls*; *hot* alone with *weather, air* or *water*; and *dog* alone with *bark* or *tail*. Another group of idioms with exactly the same collocations as one of the parts is found. An example of this would be the phrase *versed in*, where *versed* never occurs without *in*. This led Berry-Rogghe (1974, pp. 21–2) to propose the following definition of a phrasal verb:

> *Those combinations of verb + particle are to be considered as constituting a single lexical item when they contract different collocational relations from those of the particle as a separate entity.*

To identify idiomatic phrasal verbs in which the particle was *in*, Berry-Rogghe first found the right-hand collocates of *in*. Only those cases where *in* was immediately preceded by a punctuation mark were counted, lest any preceding verb impose its own collocational pattern. The significant right-hand collocates of *in*, listed in order of decreasing *z* score, were as follows: *spring, spite, short, reality, afternoon, fact, daytime, vain, Russia, summer, manner, case, order, morning, sense, world, London, country, voice, opinion, pocket, way, garden, town, minutes, road, night, book, America, days.* This list will be referred to as list 2.

Next, Berry-Rogghe found the collocational sets of all items significantly followed by *in*, which were given in list 1. The span size was two items to the right without overstepping any sentence boundary. Thus, the collocational sets were found for the phrases *interested in, came in* and so on. For each collocational set, the number of collocates which also appeared in list 2 was determined. We will call these *a*. These were collocates of both the phrase containing *in* and the lone word *in*. The total number of members in the collocational set of the phrase will be called *b*. An R score was devised, which was given by the formula R= *a/b*. The phrase *versed in* had three significant collocates, namely *politics, history* and *Greek*. None of these were significant collocates of *in* alone. Thus, for the phrase *versed in, R* = 0/3 = 0. Similarly, for the phrase *believe in,* none of the five collocates *witchcraft, God, Jesus, Devil* or *Paradise* were significant collocates of *in* alone, giving an R score of 0/5 = 0. On the other hand, the 11 significant collocates of *live in* were as follows: *hut, house, *town, *country, *London, *room, *world, *place, family, happiness* and *ignorance*. Those six words marked with asterisks also appeared in list 2, being collocates of *in* alone. In this case R = 6/11 = 0.54. The lower the R value, the more idiomatic the phrase, since the set of collocates of the phrase are distinct from the collocates of the particle alone. Thus *versed in* and *believe in* are highly idiomatic, while *live in* has relatively low idiomaticity.

A version of Berry-Rogghe's z-score formula was used to sort collocates in the TACT (Text Analytic Computer Tools) system for text navigation and analysis, so that associationally richer term pairs appear near the top of the list. This formula compares the number of times a collocate occurs in the span surrounding the node with the number of times it would have occurred if the occurrences of the collocate were randomly distributed throughout the corpus (Lancashire 1995).

3.2.2 Collocational strength in French political tracts

Geffroy et al. (1973), in their study of lexical co-occurrences in French political tracts distributed in Paris in May 1968, produced a formula for the strength of collocations (called *C*) which took into account both the frequency of co-occurrences and the proximity of the collocates to each other. It is important when studying French co-occurrences to take into account the order of **node** or **pole** (one member of a collocational pair) and collocate. In their formula, *cfo* is the frequency observed between the collocate *F* and its pole *P* within a given span, and *d* is the number of items interspersed between the pole and each collocated item. *f(P)* is the frequency of the pole. The full formula is as follows:

$$C = \frac{cfo(F,P)\sqrt{\sum(1/d)}}{f(P)}$$

For example, *travailleurs* and *étudiants* occur together in the following two sentences: *Avec les flics les travailleurs(1) et les étudiants(0) savent s'y prendre*, and *On vit de jeunes travailleurs(4) commencer(3) à venir(2) se battre(1) avec les étudiants(0)*. Thus, *travailleurs* appears in the left-side spans of *étudiants* with the coefficient

$$C = (2/2)\sqrt{(1/1) + (1/4)} = 1.07$$

Words in a corpus have a relative frequency (denoted *fr*) which is the ratio between their number of occurrences and the total number of words in the corpus; for a span of *n* items to the left and right of the pole word *P*, the quantity $n \times fr(F)$ is the theoretical co-frequency (*cft*) of *F* with *P*. This value was compared with the actual co-occurrence frequency (*cfo*) observed for *F* in the vicinity of *P*, to yield $d = cfo - cft$. Geffroy et al. apply the *z* score to this difference to determine whether their observed collocations were significant. All significant collocations can be displayed on a flat lexicograph, which lists every keyword accompanied by all significant collocates to the left and right. These collocates are called first-order collocates, and there is a direct relation between the keyword and collocate. For second-order collocates, the collocate and keyword are only indirectly related, by the fact that there is a first-order collocation between the keyword and another word, and a first-order collocation between that other word and the collocate. An example of a flat lexicograph is shown in Table 4.6, which shows that *pouvoir* always comes before the keyword *travailleurs*, *grève* always comes after it, and *lutte* co-occurs both before and after the keyword.

pouvoir	travailleurs	étudiants
revendications		grève
lutte		lutte
jeunes		immigrés
millions		français

Table 4.6 Flat lexicograph for the word *travailleurs*

The calculations carried out for a pole can in turn be made for each of the selected co-occurrents giving rise to a network depicting second- and higher-order collocations. So that only the major collocations would be depicted, Geffroy et al. employed cut-off points for minimum values of *C* and *f* such as 5 and 20. They thus produced a multistorey lexicograph, which was not simply a tree since it incorporated cycles and loops. One circuit in the diagram connected the three words *parti*, *communiste* and *français*. Different lexicographs were obtained from different sets of tracts or subcorpora, allegedly reflecting the different philosophies of communists and anarchist-Maoists.

3.2.3 The use of combinatorics to determine collocational significance

Lafon (1984) gives a method of finding the significance of a collocation, using the mathematics of combinations. Consider the following sequence of pseudowords in a text: $S\ S\ F\ G\ S\ F\ F\ G\ S\ S\ F\ G\ F\ S\ S\ S\ F\ S\ G\ F\ S\ S$. We wish to know whether F and G occur together more commonly than would be expected by chance. Lafon uses the term 'couple' to refer to a sequence of two words where the order is fixed, for example, F followed by G, and the term 'pair' to refer to two words which appear together in either order. In the above example, there are three occurrences of the couple $(F{\rightarrow}G)$, two occurrences of the couple $(G{\rightarrow}F)$, and thus five occurrences of the pair (F,G). To find the probability of obtaining these values, we use the following formula:

$$\mathrm{Prob}((F,G)=k)=\frac{\binom{f}{k}\binom{s+g}{g-k}}{\binom{f+g+s}{g}}$$

k is the number of times the pair (F,G) occurs (5), while f, g, and s are the number of times the words F, G and S each occur in the sequence (7, 4 and 11 respectively). The notation

$$\binom{n}{r}$$

means the number of ways (or combinations) r words can be selected from a sequence of n words without regard to order. Thus, if our word sequence is X, Y, Z, we can select two different words from it in three different ways: XY, YZ or XZ. Since word order is not important here, XY and YX count as one combination. The number of distinct combinations selecting r words out of n is

$$\binom{n}{r}=\frac{n!}{(n-r)!\,r!}$$

The notation $n!$ means that the value n is multiplied by $n-1$, then by $n-2$, and so on until we reach 1. In this example $n = 3$, so $n! = 3 \times 2 \times 1 = 6$. By convention, $0!$ always equals 1. In the example where $s = 11, f = 7$ and $g = 4$, probability values for all possible values of k are given in Table 4.7.

k	$Prob((F,G)=k)$	$Prob((F,G)>=k)$
$(F,G)=0$	0.187	1
$(F,G)=1$	0.435	0.813
$(F,G)=2$	0.301	0.378
$(F,G)=3$	0.072	0.077
$(F,G)=4$	0.005	0.005

Table 4.7 Probability values for different numbers of pairs occurring in a sequence

From Table 4.7 we see that the probability of the pair (F,G) occurring four or more times in a random sequence consisting of seven Fs, four Gs and 11 Ss is 0.005, so we can be 99.5 per cent confident that the collocation between F and G is a true one.

3.2.4 Daille's approach to monolingual terminology extraction

Daille (1995) presents a method for extracting technical terms from a corpus by combining linguistic data and statistical scores. Initially candidate terms are selected, which conform to one of a number of syntactic patterns encoded in **finite state automata**. Statistical scores are then used to estimate which of these candidate terms are most likely to be true technical terms.

Technical terms may be regarded as multi-word units (MWUS), since they are often composed of several words such as *receiving antenna*. The length of an MWU is defined as the number of main items (such as nouns or adjectives) it contains. Terms of length 1 are single or hyphenated terms, and those of length 2 are referred to as 'base MWUS', since they are by far the most common. In addition, most longer MWUS are built from MWUS of length 2 by the processes of com-position (e.g., [*side lobe*] *regrowth*), modification (e.g., *interfering* [*earth station*]) or co-ordination (e.g., *packet assembly/disassembly*). Both Daille (1995) and Gaussier and Langé (1994) concentrate on base MWUS of length 2. Most of these are noun phrases, obeying precise rules of syntactic combination. Examples of such syntactic patterns in French are:

- Noun Adjective, e.g., *station terrienne* (*earth station*)
- Noun1 Preposition (Determiner) Noun2, e.g., *zone de couverture* (*coverage zone*), *réflecteur à grille* (*grid reflector*) or *liaison par satellite* (*satellite link*)
- Noun1 Noun2 , e.g., *diode tunnel* (*tunnel diode*).

Examples in English are:

- Adjective Noun, e.g., *multiple access* (*accès multiple*)
- Noun2 Noun1, e.g., *data transmission* (*transmission de données*).

To detect candidate terms of length 2, sequences of words which produce regular syntactic patterns known to be typical of technical terms are found using finite state automata. A finite state automaton is an abstract mathematical formalism that receives a string of symbols as input, reads the string one symbol at a time and after reading the last symbol halts and signifies either acceptance or rejection of the input. At any stage of the computation, the finite state automaton is in one of a finite number of states. As each symbol is read in, there are rules for deciding which state the system will change to, or whether it will remain in the same state. The automaton always starts in the state known as the initial state, and if it ends up in one of the states known as a final state after reading the last symbol of the input string, the input is accepted, otherwise it is rejected. A simple automaton to test for the sequence Noun Preposition (Determiner) Noun is shown in Figure 4.1. Two possible paths may be taken through this automaton: one corresponding to the sequence Noun Preposition Determiner Noun and the other corresponding to the sequence Noun Preposition Noun. All other sequences with be rejected by this automaton.

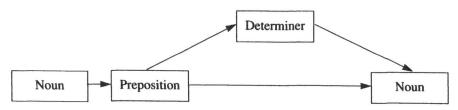

Figure 4.1 A simple finite state automaton

The syntactic patterns accepted by Daille's automata are

- Noun Adjective
- Noun1 'de' (Determiner) Noun2
- Noun1 'à' (Determiner) Noun2
- Noun1 Other-Preposition Noun2
- Noun1 Noun2.

In this way Daille was able to filter out, in terms of their syntactic structures, word co-occurrences that were possible technical terms. Statistical scores were then used in order to determine which of these co-occurrences were true technical terms. A variety of scores were evaluated to see which most successfully assigned high scores to true technical terms and low scores to co-occurrences which were not technical terms. The various statistical scores depend on the computation of three types of numeric characteristics: frequencies, **association criteria** and **Shannon diversity,** which are all described in this section. To these numeric characteristics was added a measure which uses bilingual data, namely Gaussier and Langé's affinity, which will be described in Section 3.2.6.

Each lemma pair consists of lemmas denoted Li and Lj. The frequencies needed as parameters of the association criteria are the elements of the contingency table, described in Chapter 1, Section 4.1, where:

a stands for the frequency of lemma pairs involving both Li and Lj
b stands for the frequency of pairs involving Li but not Lj
c stands for the frequency of pairs involving Lj but not Li
d stands for the frequency of pairs involving neither Li nor Lj.

The sum $a+b+c+d$, denoted N, is the total number of occurrences of all pairs obtained for a given syntactic pattern.

The association criteria and affinity compute the strength of the bond between the two lemmas of a pair, enabling candidate terms to be sorted from the most tightly to the least tightly bound. Some of these association criteria have been used in lexical statistics, such as the simple matching coefficient, the phi coefficient, the log-likelihood coefficient, and mutual information. Some have been used in other technical domains, such as biology: the Kulczinsky coefficient, the Ochiai coefficient, the Fager and McGowan coefficient, the

Yule coefficient and the McConnoughy coefficient. Daille (1995) introduces two new formulae based on mutual information, the squared and cubic association ratios. The association criteria used by Daille are as follows:

Simple matching coefficient (SMC), which varies from 0 to 1:

$$SMC = \frac{a+d}{a+b+c+d}$$

Kulczinsky coefficient (KUC), which varies from 0 to 1:

$$KUC = \frac{a}{2}\left(\frac{1}{a+b} + \frac{1}{a+c}\right)$$

Ochiai coefficient (OCH), which varies from 0 to 1:

$$OCH = \frac{a}{\sqrt{(a+b)(a+c)}}$$

Fager and McGowan coefficient (FAG), which varies from minus infinity to 1:

$$FAG = \frac{a}{\sqrt{(a+b)(a+c)}} - \frac{1}{2\sqrt{(a+b)}}$$

Yule coefficient (YUL), which varies from −1 to +1:

$$YUL = \frac{ad-bc}{ad+bc}$$

McConnoughy coefficient (MCC), which varies from −1 to +1:

$$MCC = \frac{a^2 - bc}{(a+b)(a+c)}$$

Since MCC = (2 x KUL) − 1, these two measures will produce identical rankings for the strength of association between the members of a lemma pair. For this reason, only the KUL score was evaluated by Daille.

Phi-squared coefficient (Φ^2), which varies from 0 to infinity. This score has been used by Gale and Church (1991) to align words inside aligned sentences.

$$\Phi^2 = \frac{(ad-bc)^2}{(a+b)(a+c)(b+c)(b+d)}$$

Specific mutual information (MI) varies in the range minus infinity to plus infinity. It has been used by Brown et al. (1988) for the extraction of bilingual resources and by Church and Hanks (1990) for monolingual extraction. MI may be expressed in terms of the contingency table frequencies as follows:

$$MI = \log_2 \frac{aN}{(a+b)(a+c)}$$

This measure gives too much weight to rare events. For example, if we have a word which occurs just once in the first language and another which occurs just once in the second, and they both occur within the same aligned region, the mutual information will be higher than if the words each appeared twice and both times occurred in aligned regions. In order to give more weight to frequent events, the a on the top line of the MI formula was successively replaced by all powers of a from two to 10. The cube of a was empirically found to be the most effective coefficient, yielding the cubic association ratio (MI3)

$$MI3 = \log_2 \frac{a^3 N}{(a+b)(a+c)}$$

Log–likelihood coefficient (LL), introduced by Dunning (1993):

$$LL = 2 \times (a \log a + b \log b + c \log c + d \log d$$
$$-(a+b) \log(a+b) - (a+c)\log(a+c)$$
$$-(b+d)\log(b+d) - (c+d)\log(c+d)$$
$$+ (a+b+c+d)\log(a+b+c+d))$$

where each log is to the base e.

Daille also used Shannon diversity to discriminate between candidate technical terms. In theory, a lemma which appears in various different lemma pairs is an item which either very frequently allows the creation of a term (such as *système*) or never leads to the creation of a new term (such as *caractéristique*). Consider a small French technical corpus, where just three nouns and three adjectives have been found. The frequencies of each possible lemma pair in the format (*noun-i, adjective-j*) are shown in Table 4.8. For example, the frequency of the lemma pair (*onde, circulaire*), denoted *nb(onde, circulaire)* is found at the intersection of the row for the noun *onde* and the column for the adjective *circulaire*. We are particularly interested in the 'marginal totals', *nb(i,-)* and *nb(-,j)*, such as *nb(onde,-)* which is the total number of lemma pairs in which *onde* was the first term and *nb(-,porteur)*, which is the number of lemma pairs in which *porteur* was the second term.

N-i, Adj-j	progressif	circulaire	porteur	Total
onde	19	4	6	nb(onde,-)
limiteur	9	0	0	nb(limiteur,-)
cornet	0	2	0	nb(cornet,-)
Total	nb(-,progressif)	nb(-,circulaire)	nb(-,porteur)	

Table 4.8 Relative frequencies of lemma pairs in a hypothetical corpus

Using the values in Table 4.8 (which is the contingency table of the *noun-i, adjective-j* structure), Shannon's diversity of the noun *onde* is equal to:

$H(onde,-) = nb(onde,-).\log nb(onde,-) - (nb(onde,progressif)).\log nb(onde,progressif)$
$+ nb(onde,circulaire).\log nb(onde,circulaire) + nb(onde,porteur).\log nb(onde,porteur))$.

Once again all the logs are to the base e.

Expressed formally,

$$H(i,-) = nb(i,-) \log nb(i,-) - \sum_j nb(i,j) \log nb(i,j)$$

and analogously,

$$H(-,j) = nb(-,j) \log nb(-,j) - \sum_i nb(i,j) \log nb(i,j)$$

Two related quantities, hi and hj, were found, by dividing the diversity values $H(i,-)$ and $H(-,j)$ respectively by nij, the number of occurrences of the pairs.

Daille performed a graphical evaluation of all the measures discussed in this section. Candidate terms were counted as true technical terms if they either appeared in the Eurodicautom terminology data bank or two out of three domain experts judged them to be so. For each measure, the candidate terms were sorted according to decreasing score. This list was divided into equivalence classes each containing 50 successive pairs. The results of each score were represented graphically as a histogram in which the points along the x-axis corresponded to the equivalence classes, and the y-axis was the proportion of lemma pairs which were true technical terms. For an ideal measure, the histogram would show that all the lemma pairs in the topmost equivalence classes were true technical terms, while none of the lemma pairs in the bottommost equivalence classes were true technical terms. Between these two types of equivalence classes, there would be a single sharp cut-off point or threshold. The histograms for 18 different measures were found, and compared with the ideal histogram. The histogram for MI3 and the ideal histogram are shown in Figure 4.2.

Ideal histogram

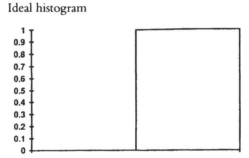

Histogram for MI3

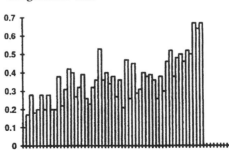

Figure 4.2 Comparison between MI3 and an ideal measure

The measures examined were $N1, N2, NC, \Phi^2, OCH, YUL, FAG, AFF, hl, h2, H(i,-),$ $H(-,j), LL, MI, MI2, MI3, KUC$ and SMC. $N1$ is the number of times the first lemma appears first in all lemma pairs, $N2$ is the number of times the second lemma appears second in all lemma pairs, NC is simple frequency of the lemma pair, AFF is the affinity measure of Gaussier and Langé (1994), described in Section 3.2.6, and $MI2$ is the square association ratio, analogous to the cubic association ratio.

The only four measures which seemed at all effective in the identification of true technical terms were the log-likelihood coefficient, the Fager and McGowan coefficient, the cubic association ratio and the simple frequency of the lemma pair. Of these four, the log-likelihood coefficient was preferred, since it is a well-established statistical technique, adequately takes into account the frequency of the lemma pairs and behaves well whatever the corpus size. Based on the shape of the histogram alone, simple frequency would be the best measure, but by definition it cannot pick out any low-frequency technical terms.

3.2.5 Mutual information for the extraction of bilingual word pairs

Gaussier, Langé and Meunier (1992) used the specific mutual information measure to automatically extract bilingual word couples from bilingual corpora. They start with the Canadian Hansards, a bilingual corpus in both English and French, where each English source sentence is aligned with its target French translation. The strength of association between an English word denoted e and a French word denoted f will be high if these two words are translations of each other. The formula for specific mutual information, $I(e,f)$, as used by Gaussier, Langé and Meunier, is as follows:

$$I(e,f) = \log_2 \frac{p(e,f)}{p(e)p(f)}$$

In this context, $p(e,f)$ is the probability of finding both e and f in aligned sentences. To find $p(e,f)$, the number of occasions e and f are found in aligned sentences within the corpus is divided by the total number of aligned sentences in the corpus, whether they contain e or f or not. $p(e)$ is the probability of finding e in an English sentence (found by dividing the number of times e occurs in the corpus by the total number of aligned sentences in the corpus) and $p(f)$ is the probability of finding f in a French sentence.

In order to use specific mutual information in this way to find which pairs of words are translations of each other, the corpus must first be aligned. Gaussier, Langé and Meunier improved the alignment program of Brown, Lai and Mercer (1991) by taking advantage of textual mark-up such as SGML and using additional lexical information such as **transwords**. These are words which remain unchanged in the process of translation, in particular numerics or wholly upper-case words, such as product names or computer commands.

The initial list of candidate translations for an English word consisted of all words which occur in at least one French sentence aligned with a sentence containing the English word. The mutual information between each candidate French word and the English word is calculated, and those French words having mutual information greater than an empirically found threshold are retained. For example, the following list of candidate translations for the English word *prime* was found, and is given in Table 4.9, along with the observed mutual information.

French word	Mutual information
sein	5.63
bureau	5.63
trudeau	5.34
premier	5.25
résidence	5.12
intention	4.57
no	4.53
session	4.34

Table 4.9 French words with highest mutual information with the English word *prime*

Using their best match criterion, Gaussier et al. eliminated from Table 4.9 all words which had been found to have a higher mutual information score with an English word other than *prime*. This left only the word *premier*, which was deemed to be the French translation of *prime*. In their experiments, about 65 per cent of English words were assigned their correct French translations, about 25 per cent had no French word assigned (these were mainly words with no real French equivalent) and about 10 per cent had an incorrect French translation assigned to them.

3.2.6 Pattern affinities and the extraction of bilingual terminology

There is a need for multilingual terminology extraction, both for the documentation of products and the standardisation of terminology. As industrial activities become increasingly international, there is an increased need for multilingual terminology. Acquisition of such terminology can be performed by domain experts, but is a slow and expensive process. Gaussier and Langé (1994) address a method of partially automating bilingual term acquisition, to be discussed below.

After Daille, a finite state machine was implemented which extracts MWU candidates in the form of lemma pairs. These candidates are not always 'good' terms but of syntactic patterns known to be productive. About 4000 candidates were found for each language in a part-of-speech-tagged and lemmatised bilingual corpus of about 200,000 words. The corpus had been manually aligned at the sentence level, and was used to derive measures of association between the candidate terms in the English side of the corpus and their possible French translations on the other side, based on the number of times a source candidate and a target candidate occurred in aligned sentences. Gaussier and Langé tested four measures based on contingency tables: mutual information, phi-square, Fager and McGowan's coefficient and the log-likelihood coefficient. A linguistically motivated improvement was tried, based on the observation that in 81 per cent of cases, a French base MWU of the form *Noun1 de Noun2* was a translation of an English base MWU of the form *Noun2 Noun1*. It was proposed that the association measures between term pairs be enhanced by a weighting factor based on pattern affinity, which was the probability that the syntactic pattern of the source candidate gets translated into the syntactic pattern of the target candidate.

The second method made use of bilingual associations between the single words composing the candidates to derive a measure between the candidates. For example, the mutual information between the English candidate term *earth station* and the French *station terrienne* was the sum of the mutual information between the first English word and the first French word (*earth* and *station*), the first English word and the second French word (*earth* and *terrienne*), the second English word and the first French word (*station* and *station*), and the second English word and the second French word (*station* and *terrienne*). This measure was called **mutual information with double association**.

In order to evaluate these measures, Gaussier and Langé produced a reference list containing (source, target) pairs of base MWUs. This list was made from an existing list established by terminologists, complemented with good pairs extracted from the corpus. Although this list was not definitive, it allowed a fast, costless evaluation of different approaches to term extraction. The candidate list was automatically sorted according to decreasing association score and compared with the reference list. In order to perform an evaluation, a list of the *n* best candidates according to a certain score was compared with the reference list, according to the criteria of recall and precision. In many information retrieval applications, retrieval effectiveness is measured according to these two criteria (Salton and McGill 1983). **Recall** measures the proportion of relevant information retrieved in response to a search procedure (the number of relevant items actually obtained divided by the total number which would have been obtained in a perfect search). **Precision** measures the proportion of retrieved items that are in fact relevant (the number of relevant items obtained divided by the total number of retrieved items). Candidate terms were considered to be good if since they occurred in the reference list. Precision and recall were then found using the following formulae:

$$\text{Precision} = \frac{\text{Number of good candidates}}{\text{Total number of candidates}}$$

$$\text{Recall} = \frac{\text{Number of good candidates}}{\text{Number of terms in the reference list}}$$

The following measures were compared: log-likelihood, MI, MI with pattern affinity weight, and MI with double association. It was found that MI with pattern affinities gave slightly better results than MI alone. For example, for the 100 highest-scoring candidates in each list, MI alone gave 90 per cent precision and 7.2 per cent recall, while MI with pattern affinity gave 94 per cent precision and 7.6 per cent recall. Using the first method, the log-likelihood measure was found to degrade rapidly as the length of the candidate list was increased, but, unlike mutual information, did not privilege rare events. MI was preferred, however, since it allows the incorporation of pattern affinities. The second method (double association) was generally better than the first, giving good recall especially for long candidate lists. However, the two different methods emphasise quite different associations. Of the first 100 candidates obtained by each method, only six overlap. Thus, it is best to combine the candidates from both measures. In this, a list of mainly good terms was prepared and shown to a terminologist. It would be a much simpler task simply to delete the poor candidates than to create the entire list manually. The overall satisfaction of the human post-editor can be improved by setting the affinity measure thresholds in order to discard less relevant candidates.

3.2.7 Kay and Röscheisen's text–translation alignment algorithm

Most techniques which use measures of affinity between words across languages depend on the prior alignment of parallel texts. We will thus devote the next two sections to descriptions of automatic text alignment algorithms. The corpora are divided into regions, typically one or two sentences long, to show which sentence or sentences of one language correspond with which sentence or sentences in the other. The two languages are referred to as the source language and the target language. In machine translation, we say that we are translating from the source language to the target language. According to Kay and Röscheisen (1993), parallel aligned corpora are useful as an aid to translators, students of translation, designers of translation systems and lexicographers. The advantages of using aligned corpora as a source of lexicographic data as opposed to published dictionaries or term banks are that one can obtain information on topical matters of intense though transitory interest, and that, with contemporary corpus data, one can obtain information on recently coined terms in the target language.

The alignment procedure of Kay and Röscheisen depends on no information about the language pair involved other than what can be derived from the texts themselves. The method is based on the premise that a pair of sentences containing an aligned pair of words must themselves be aligned, and requires the establishment of a word sentence index (WSI), an alignable sentence table (AST), a word alignment table (WAT) and a sentence alignment table, described below.

The WSI is a table with an entry for each different word in the text, showing the sentences in which that word occurs. For example, the entry *proton 5 15* would show that the word *proton* occurs in sentences 5 and 15. One WSI is required for the text in each language.

The AST is a matrix where each entry (m,n) shows whether or not the *m*th sentence of the source language and the *n*th sentence of the target language could possibly be aligned. Some alignments are forbidden, because of the following constraints:

1. the first sentence of the source language must align with the first sentence of the target language
2. the last sentence of the source language must align with the last sentence of the target language
3. there must be alignment whenever an 'anchor' point has been identified, which is where a word in one language is without doubt a translation of a word in the other. For all other points, the range of possible target sentences with which a source sentence can align increases with the distance from an anchor point.

The WAT is a list of pairs of words, one in each language, which have been found using Dice's coefficient to have similar distributions in the text, and thus can be confidently assumed to be translations of each other. Dice's coefficient is found using the formula $2c/(a + b)$, where c is the number of times the two words occur in corresponding positions, a is the number of times the source word occurs in the text and b is the number of times the target word appears in the text. For example, if the word *electric* occurs in sentences 50, 62, 75 and 200 of an English text, and the word *elektrisch* occurs in sentences 40 and 180 of a German text, and the AST contains the pairs <50, 40> and <200, 180>, but none of the other possible pairings of these words, then Dice's coefficient is (2 x 2)/(4+2) = 0.67. For inclusion in the WAT, a word pair must have an above threshold similarity metric and must also occur at least a given number of times in the corpus.

The sentence alignment table (SAT) records for each pair of sentences how many times the two sentences were set in correspondence by the following procedure, where sentence pairs are associated if they contain words that are paired in the WAT. For each word pair v, w a correspondence set is created. A sentence pair <*sv,sw*>, containing words v and w, is included in the set if that word pair is in the AST, v occurs in no other sentence *sh*, and w occurs in no other sentence *sg*, such that <*sv,sh*> or <*sw,sg*> is also in the AST. Each sentence pair in the correspondence set of the word pair <*v,w*> is added to the SAT. A count is kept of the number of times each particular sentence pair association is supported. Sentence associations that are supported more than a threshold number of times are transferred to the AST.

The process continues in cyclical fashion, where the updated AST is used to compute a new WAT, which in turn is used to produce a new SAT. The procedure stops after a certain number of iterations, or when no further updates are produced in a whole iteration. Throughout the procedure, no crossover in the sentence alignment is allowed. That is to say, if two sentences *sv* and *sw* are aligned, then sentences occurring before *sv* cannot be aligned with sentences occurring after *sw*. The SAT is the final output of the program.

3.2.8 Production of probabilistic dictionaries from pre-aligned corpora

The Kay and Röscheisen (1993) algorithm performs two functions simultaneously, namely sentence alignment as recorded in the sentence alignment table, and word alignment as recorded in the word alignment table. This is, in effect, a probabilistic dictionary. Other algorithms such as that of Brown, Lai and Mercer (1991) or Gale and Church (1993), described in Chapter 3, Section 4.9.3, also perform the task of sentence alignment, thus producing parallel aligned corpora, but other statistics such as those described by Daille (1995) must be used if probabilistic dictionaries are to be created from parallel aligned corpora. In order to create bilingual word lists from parallel sections of the International Telecommunications Union (ITU) corpus, McEnery and Oakes (1996) concentrated on the identification of words in English which had many characters in common with their candidate French translations. To quantify the degree of lexical similarity between such word pairs, they used approximate string matching techniques such as those described in Chapter 3, Section 4, and in particular, Dice's similarity coefficient. They were able to demonstrate that prior alignment of the corpus greatly enhanced the accuracy of identifying bilingual word pairs using approximate string matching techniques. First of all, they used word pairs drawn from the respective vocabulary lists of the unaligned corpus. Thus, lexically similar word pairs were considered to be translations of each other, even though there was no guarantee that they arose from parallel sections of the corpus. Their results for this part of their experiment are given in Table 4.10. In each case, the accuracy value (Accuracy (a)) given is the percentage of lexically similar word pairs which were true translations of each other. Their results are also shown (Accuracy (b)) for the aligned corpus, where the identified word pairs were both lexically similar and arose from corresponding sentences of the corpus. It may be seen that the accuracy of their method was greatly enhanced by prior alignment of the corpus.

Dice score	0.4	0.5	0.6	0.7	0.8	0.9	1
Accuracy(a)	0	7	21.5	49	81.6	97	100
Accuracy(b)	41.1	92.7	90.2	100	100	100	100

Table 4.10 Accuracy of Dice's similarity coefficient in retrieving bilingual word pairs from (a) the token list of an unaligned corpus and (b) parallel regions of an aligned corpus

3.2.9 Collocations involving more than two words: the cost criterion

In order to extract collocations from corpora, Kita et al. (1994) used their 'cost criterion', which depends on both the absolute frequency of collocations and their length in words. Absolute frequency alone is not an effective measure for comparing overlapping phrases such as *in spite* and *in spite of*, because the frequency of the shorter sequence will always be more than or equal to that of the longer sequence. However, the word *of* occurs so often immediately after *in spite*, that we must conclude that *in spite of* is the full collocation rather than *in spite*. This observation may be expressed using the concept of 'reduced cost'. In our example, *a* is the subsequence *in spite*, *b* is the sequence *in spite of*, *f(a)* is the frequency of *in spite* in the corpus, *f(b)* is the frequency of *in spite of* and |*a*| is the length in words of *in spite* (2). The reduced cost for *a*, denoted *K(a)*, is given by the formula

$$K(a) = (\ |a| - 1\ ) \times (f(a) - f(b))$$

Thus, since *in spite* is almost always followed by *of*, *f(a)* and *f(b)* will be almost the same and the value of *K(a)* will be low. However, if we make *in spite of* equal to *a*, and *in spite of everything* equal to *b*, the frequency of *b* will be less than that of *a*, since *in spite of* can be followed by a number of fairly likely possibilities, such as *that*, *this* or *everything*. In this case, *K(a)* will be greater than it was for *in spite*, suggesting that *in spite of* is more likely to be the full collocation.

Kita et al. used both mutual information and the cost criterion to extract collocations from the ADD (ATR Dialogue Database, consisting of parallel keyboard and telephone conversations in Japanese and English). The subsection of the corpus they used concerned travel information. The cost criterion was found to be the more suitable measure for language learning purposes, picking out everyday phrases such as *is that so*, *thank you very much* and *I would like to*, while mutual information tended to pick out specific terms such as *Fifth Avenue* or *slide projector*.

Jelinek (1990) suggests a generalisation of the mutual information formula which allows one to find the associative strength of collocations involving more than two words. First of all the mutual information is found for pairs of words, and those pairs with above-threshold mutual information are retained. At the next round of this iterative procedure, these pairs are treated as single words, and the mutual information between these pairs treated as single words and other single words is found. The resulting pairs with above-threshold mutual information from this iteration are then regarded as single items for the next iteration. This procedure continues, producing ever-longer word sequences, until one iteration takes place in which no sequence-word pairs are found with above-threshold mutual information.

3.2.10 Factor analysis to show co-occurrence patterns among collocations

A knowledge of collocations can show how many different senses a word has, since each different sense of a word will have its own set of collocates. Biber

(1993) uses factor analysis to find different sets of collocates for an ambiguous word in a monolingual corpus, as will be described in this section, while Gale, Church and Yarowsky (1992) use the collocations between the different senses of a word and their possible translations in a bilingual parallel corpus, as will be described in the next section. Biber states that one of the main problems for applied natural language processing are gaps in the lexicon, including missing words and word senses. The lexicon can be enhanced relatively easily due to the availability of electronically-stored corpora, but the identification of additional word senses is a more difficult task. It is often difficult to manually group concordance entries according to word sense, since there may be a vast number of entries for the word of interest. As we have seen, statistical techniques such as the use of mutual information or t scores (Church et al. 1991) have been used to measure the differences between the collocational behaviour of near-synonyms such as *strong* and *powerful*. However, Biber points out that these methods alone do not show how many different senses a word has, and thus human judgement must also be used.

Biber used the technique of factor analysis to identify the basic word senses associated with the words *certain* and *right*. This technique, described in more detail in Chapter 3, Sections 2.6 and 2.7, shows which groups of words tend to co-occur in different subtexts of a corpus. If each of these subtexts coherently focuses on a single topic, then the different word groupings containing the word of interest that appear in these subtexts will often reveal the different possible senses of that word, since words will be used in a single sense throughout the domain of a subtext. The input data for Biber's factor analysis consisted of the frequency counts of each collocation of a given word with *certain* or *right* which occurred more than 30 times in an 11 million-word subsample of the Lancaster–Longman corpus. It was assumed that each of these collocational pairs would have a strong relation to a single sense of *certain* or *right*, and it was hypothesised that the collocational pairs corresponding to a given sense of the word under study would be grouped together by the factor analysis. Each collocational pair was associated with each resulting word group or 'factor' to a degree measured by its 'loading'. Only the larger loadings were considered in interpreting the factors. Three factors were found for *certain*, the first two factors grouping word pairs which indicated that *certain* was used in the sense of *particular*, and the third factor grouping word pairs where *certain* was used in the sense of *sure*. The first factor included the pairs *certain other* and *certain types*, the second included *certain extent* and *certain aspects*, while the third included *certain that* and *make certain*. Four factors were found for *right*, corresponding to the senses of the right hand side, directly or exactly, ok or correct, and a less clear-cut category. The first factor included the pairs *right hemisphere* and *right ear*, the second included *right there* and *right back*, the third *all right* and *that's right* and the fourth *right you* and *right so*.

3.2.11 Mutual information for word segmentation

A slightly different grouping problem faces us when we consider Chinese. Chinese words may consist of a single character, but often two or more characters are required to constitute a single word. In Western languages, lexical words are delimited by spaces, but this is not the case in Chinese texts and so a method of segmenting text into words other than the recognition of spaces is required. Luk (1994) proposes word segmentation based on the bigram technique which employs a dictionary of bigrams or two-character sequences extracted from the text which have either a high frequency of co-occurrence or high specific mutual information. A segmentation marker is placed between two adjacent characters in the text if this two-character sequence does not appear in the dictionary. This method effectively extends the concept of mutual information to cover sequences of three or more characters. Using this method, the overall mutual information of a sequence is taken to be the mutual information of its 'weakest link' or the two-character subsequence with the lowest mutual information of any adjacent character pair in the sequence.

3.2.12 Smadja's XTRACT program for the extraction of collocations

Smajda (1991, 1992) describes the XTRACT[5] program which was designed to automatically extract collocations from a corpus. It is not a concordance program as such, but it does represent the type of processing which would be good for a concordancer to carry out. The domain of the corpus used by Smadja was newswire stories from the Associated Press, where collocations are very common. In the news story he analysed (1992), every sentence contained at least one collocation. As collocations found in a corpus belonging to a particular genre such as news stories are sublanguage specific it is of particular value to extract collocations from them, since compiled lists of collocations used in a particular sublanguage are generally unavailable.

Smadja assumes that two words co-occur if they appear in the same sentence and are separated by less than five words. For each pair of collocating words, a vector of 10 values is created, such as that shown in Table 4.11. Each value corresponds to the number of times the two words were found p words away from each other. For example, the entry under p-3 for the words *trade* and *free* shows the number of times the second word *free* was found to occur exactly three words before the first word *trade*. In a **flexible collocation**, the words may be inflected, the word order may vary and the words can be separated by any number of intervening words. For example, the phrase *to take steps* can appear as *took immediate steps* or *steps were taken to*. If the word order and inflections are fixed, and the words are in sequence, the collocation is said to be **rigid**. An example of a rigid collocation is a fixed compound such as *International Human Rights Covenants*. The collocation between the two words of Table 4.11 is rigid, since 99 per cent of the instances are in the form *free trade*. There are no instances of *trade free*, an example of the two words being in complementary distribution. The other values in the table are similar to those expected if the co-occurrence between the two words were more or less random.

Total	$p-5$	$p-4$	$p-3$	$p-2$	$p-1$	$p+1$	$p+2$	$p+3$	$p+4$	$p+5$
8031	7	6	13	5	7918	0	12	20	26	24

Table 4.11 Associations between *trade* and *free*

The goal of the first stage of the XTRACT program is to isolate frequent and strong relationships such as *free trade*. A table similar to Table 4.11 is created for every word pair in the corpus, and the part of speech of each collocate is recorded. An optional feature of XTRACT is that all the words involved in lexical relations can be converted to their inflectional lemmas, when the data for words which map to a common lemma is combined. It is assumed that if two words appear in a collocation they must not only appear together significantly more often than expected by chance, but also in a relatively rigid way. Thus, the first filtering process used by XTRACT to extract the most likely collocations is selecting the collocates of a given word with a large total frequency (found by summing the values for all 10 positions in Table 4.11.

The second filtering process examines the statistical distribution of the positional occurrences. Word pairs producing a flat histogram (where the height of the bars depicts the number of occurrences and each bar corresponds to a relative position) show that the frequency of occurrence is largely independent of relative spacing, so the collocation is not rigid. Word pairs are thus preferred if they produce a histogram with a distinct peak or peaks, since the greater the height, the more strongly the two words are lexically related. Lexical relations are filtered if the height is below an empirically found threshold. A second criterion used in the filtering process is that of 'spread', which is the variance for the frequency values of each collocate at each of the possible 10 positions. Retaining only those word pairs which show high spread enables the removal of 'conceptual collocations' such as *bomb–soldier* and *trouble–problem*, collocations in which the two words co–occur because they belong in the same context rather than form a true lexical collocation (Smadja 1991).

The third filtering process uses both statistical and part of speech information. A collocation is accepted if the two **seed words** are consistently used in the same syntactic relation. To be retained, a word pair must consist of the same syntactic units (such as verb-object, verb-subject, noun-adjective or noun-noun) at least on a threshold percentage of occasions on which it appears. In the following section, we will see how a program called *Champollion* is able to start with the pairwise associations produced by XTRACT, and from these produce collocations involving more than two words.

3.2.13 Extraction of multi-word collocation units by Champollion

Smadja, McKeown and Hatzivassiloglou (1996) have produced a program called *Champollion*, which, if supplied with a parallel corpus in two different languages and a list of collocations in one of them, uses statistical methods to automatically produce the translations of those collocations. The XTRACT program described in Section 3.2.12 was used with the English portion of the Canadian Hansard corpus to compile a list of English collocations. The task of translation itself requires a prealigned corpus, and this alignment was performed using the Gale and Church algorithm, described in Chapter 3, Section 4.9.3. Since this algorithm is most accurate when identifying 1:1 alignments (where, for example, one sentence of English corresponds exactly with one sentence of French), only those sections of the corpus which were found to align in 1:1 fashion were used, which constituted about 90 per cent of the corpus.

Once the list of source language collocations has been obtained, *Champollion* starts by identifying individual words in the target language that are highly correlated with the source collocation, using Dice's coefficient as the similarity measure. In this way a set of words in the target language is produced, from which the candidate target language collocations will be derived. It is assumed throughout that each source language collocation is both unambiguous and has a unique translation in the target language. This means that decisions about which words or phrases are translations of each other can be based solely on their mutual co-occurrence behaviour within the corpus, and the immediate contexts surrounding them can be ignored.

Dice's similarity coefficient for a source phrase X and a target phrase Y is given by the formula $Dice\ (X, Y) = 2 \times fxy/(fx + fy)$, where fxy is the number of aligned regions in which both the source phrase X and the target phrase Y appear, fx is the total number of regions in which the source phrase X appears and fy is the total number of regions in which target phrase Y appears. Throughout the procedure of finding the best translation for each source collocation, sub-optimal translations and partial translations are successively seeded out if their Dice coefficient with respect to the source collocation falls below an empirically selected value.

Dice's coefficient was considered superior to information theoretic measures such as mutual information which are widely used in computational linguistics. The formulae for both specific mutual information and average mutual information have been given in Chapter 2, Section 2.7. The main reason that Dice's coefficient was preferred was that in this translation process, an aligned region of the corpus where one or both of the source and target phrases is present is a far more significant finding than a region where neither is present. This is because such a positive match is a much rarer occurrence than a negative match. The coefficient reflects this, since only the number of times where one (fx or fy) or both of the phrases (fxy) are present are included in the formula. If using an information theoretic measure such as average mutual information, on the other hand, instances where neither phrase is present are assigned equal importance to instances where both source and target phrase are present. This is because information theory considers both possible outcomes of a binary selection process (0 or 1) to have equal precedence, and thus a positive match when two variables have the value 1 is no more or less significant than a negative match when two variables equal 0. Average mutual information is said to be a completely symmetric measure: if all the ones were exchanged for zeros and vice versa, its overall value would remain the same. With specific mutual information we have an intermediate situation, since the measure is neither completely symmetric nor does it completely ignore negative matches.

Given a source English collocation, *Champollion* first identifies in the English section of the corpus all the sentences containing the source collocation. It then attempts to find all words that can be part of the translation of the collocation; that is, those target language words that are highly correlated with the source collocation as a whole. These words satisfy the following two conditions:

1. the value of the Dice coefficient between the word and the source collocation W is at least Td, where Td is an empirically chosen threshold
2. the word appears in the target language opposite the source collocation at least Tf times, where Tf is another empirically chosen threshold.

This second criterion helps limit the size of the set of words from which the final translation will eventually be produced.

Now that this original set of target words highly correlated with the source collocation has been identified, *Champollion* iteratively combines these words into groups. The program first forms all possible pairs of words in the original set, and identifies any pairs that are highly correlated with the source collocation, using criteria analogous to (a) and (b) above, except that the correlation between pairs of words rather than individual words is found. Next, triplets are produced by adding a highly correlated word to a highly correlated pair (producing the Cartesian product of the word groups retained at the previous stage and the single words in the original highly correlated set), and the triplets that are highly correlated with the source language collocation are passed to the next stage. This is the main iteration stage of the algorithm, and the process is repeated until no more highly correlated combinations of words can be found. At each stage of the process, it is necessary to identify the locally best translation, this being the single word, word pair, word triple and so on that has the greatest Dice score with the source translation of all possible translations with that number of words. Each of these word groups is stored in a table of candidate final translations, along with its length in words and Dice score. The entries in this table for each iteration when the source collocation is *official languages* are given in Table 4.12. In each case, the number of the iteration is given first, followed by the highest-scoring word set found at that iteration, then the Dice score for that translation, and finally the number of retained word sets at that iteration. Finally, on the ninth iteration, no highly correlated word group is found and the iteration stops. Using this table, *Champollion* selects the group of words with the highest similarity coefficient as the target collocation. If two word groups have the same similarity coefficient with the source collocation, the longer is taken to be the translation.

---

1. officielles 0.94 (11)
2. officielles langues 0.95 (35)
3. honneur officielles langues 0.45 (61)
4. déposer honneur officielles langues 0.36 (71)
5. déposer pétitions honneur officielles langues 0.34 (56)
6. déposer lewis pétitions honneur officielles langues 0.32 (28)
7. doug déposer lewis pétitions honneur officielles langues 0.32 (8)
8. suivantes doug déposer lewis pétitions honneur officielles langues 0.20 (1)

Table 4.12 Candidate translations at each iteration of the *Champollion* process for the source collocation *Official Languages*

Having found the best translation among the top candidates in each group of words, the program determines whether the selected translation is a single word, a flexible collocation or a rigid collocation by examining samples in the corpus. For rigid collocations, the order in which the words of the translation appear in the corpus is reported. The definition of a rigid collocation is that in at least 60 per cent of cases the words should appear in the same order and at the same distance from one another.

Champollion does not translate closed-class words such as prepositions and articles. Their frequency is so high in comparison to open-class words that including them in the candidate translations would unduly affect the correlation metric. Whenever a translation should have included one closed-class word, the program produces a rigid collocation with an empty slot where that word should be.

3.2.14 Use of a bilingual corpus to disambiguate word sense

Collocation is not the only means available to us for disambiguating word meaning. Dagan, Itai and Schwall (1991) argued that 'two languages are more informative than one' in disambiguating word sense. They showed that it was possible to use the differences between certain languages to infer information about the different meanings of certain words. Rather than using differences between the sets of collocates for different meanings of a word as a means of discriminating between them, it is possible, using an aligned corpus, to study the co-occurrence pattern of a word in one language with its translation in the other language in aligned sections of the corpus. Such co-occurrences are similar to collocations in a single language, if we consider the allowable span in which two collocates may appear to include the aligned section of the other language. Gale, Church and Yarowsky (1992) followed this approach, making use of the Canadian Hansards which are available in machine-readable form in both English and French. For example, they considered the polysemous word *sentence*, which may mean either a judicial sentence (normally translated by *peine*) or a syntactic sentence (translated as *phrase*). This simple technique will not always work, since *interest* translates as *intérêt* for both monetary and intellectual interest. And in Japanese, for example, the word for *wear* varies according to the part of the body, whereas in English we would make no distinction.

The Hansards provided a considerable amount of training and testing material. In the training phase, Gale, Church and Yarowsky collected a number of instances of *sentence* translated as *peine*, and a number translated by *phrase*. Then, in the testing phase, the task was to assign a new instance of *sentence* to one of the two senses. This was attempted by comparing the context of the unknown instance with the contexts of known instances.

The following formula was used to sort contexts denoted *c*, where contexts were defined as a window 50 words to the left and 50 words to the right of the ambiguous word:

$$score(c) = \prod_{token\ in\ c} Pr(token|sense1) / Pr(token|sense2)$$

This formula means that for every word token in the context, we must estimate the probability that the token appears in the context of *sense*1 or *sense*2, then multiply together the resulting *Pr(token | sense*1)/*Pr(token | sense*2) ratio for every word. Gale, Church and Yarowsky found that words which have a large value for the quantity *frequency(token | sense*1) x log *Pr(token | sense*1)/*Pr(token | sense*2) tend to provide important contextual clues for the scoring of contexts. Table 4.13 shows some words that, when found in conjunction with the polysemous term *drug*, have a large value for this quantity. The actual base of the logarithms they used is not stated, but any base would give the same rank order of contexts.

Word	Sense	Contextual clues
drug	medicaments	prices, prescription, patent, increase, generic, companies, upon, consumers, higher, price, consumer, multinational, pharmaceutical, costs.
drug	drogues	abuse, paraphernalia, illicit, use, trafficking, problem, food, sale, alcohol, shops, crime, cocaine, epidemic, national, narcotic, strategy, head, control, marijuana, welfare, illegal, traffickers, controlled, fight, dogs.

Table 4.13 Contextual clues for sense disambiguation for the word *drug*

The next task was sentence alignment. Gale, Church and Yarowsky began by aligning the parallel texts at the sentence level using the method of Gale and Church (1991). The next step was to establish word correspondences, for which they used a program that is intended to identify which words in the English text correspond to which words in the French text. They use the term 'alignment' to mean that order constraints must be preserved. Thus, if English sentence *e* is aligned with French sentence *f*, sentences occurring before *e* cannot be set into correspondence with sentences occurring after *f*. Conversely, 'correspondence' means that crossing dependencies are permitted. The word correspondence task was enabled by the creation of contingency tables, as described in Section 3.2.4. Using these tables, the association between any two words can be found by making use of any one of a number of association measures such as mutual information. An example of input and output for the word alignment program is given in Table 4.14.

```
Input:

We took the initiative in accessing and amending current legislation and policies to ensure
that they reflect a broad interpretation of the charter. = Nous avons pris l'initiative
d'évaluer et de modifier des lois et des politiques en vigeur afin qu'elles correspondent
à une interprétation généreuse de la charte.


Output:

We took the initiative in assessing and amending current legislation and policies      to
    pris     initiative     évaluer     modifier       lois           politiques

ensure that they reflect     a broad     interpretation of the charter.
afin                correspondent    généreuse interpretation         charte
```

Table 4.14 Input and output of a word alignment program

The conditional probabilities *Pr(tok/sense1)* and *Pr(tok/sense2)* can be estimated in principle by examining the 100-word contexts surrounding instances of one sense of an ambiguous word such as *duty*, counting the frequency of each word appearing in those contexts, and dividing the counts by the total number of words appearing in those contexts. This is the called the maximum likelihood estimate (MLE) which has a number of problems, one of which is that zero probability will be assigned to words that do not happen to appear in the examined contexts, even though such words occur in the corpus as a whole. In fact Gale, Church and Yarowsky use a more complex method of estimation called the **interpolation procedure**, which takes into account the frequencies of words in the corpus as a whole as well as the frequency of words in the 100-word contexts of the ambiguous word. The interpolation procedure provides a weight for each word, and the product of this weight and the frequency of the word in the corpus as a whole is a measure of how much the appearance of that word supports one sense or the other of the ambiguous word. For the tax sense of the word *duty*, the four words with the highest *weight* x *frequency* product were *countervailing, duties, us* and *trade*, while words with a high product for the obligation sense of duty included *petition, honour* and *order*.

3.2.15 G-square or log likelihood in the development of a 'phrasicon'

Milton (1997) has used the G-square, G score or log likelihood measure to find overused and underused phrases in a corpus of learners' English when compared with a corpus of standard English. Examples of underused and overused phrases consisting of four units (words or punctuation marks) are shown in Table 4.15.

Underused phrases	G score	Overused phrases	G score
In this case the	12.33	First of all,	198.39
It has also been	10.96	On the other hand	178.34
It can be seen	10.96	In my opinion,	86.02
An example of this	10.96	All in all,	67.90
good example of this	9.59	In fact, the	48.08
This is not to	9.59	In addition, the	38.44
In an ideal world	9.59	In a nutshell,	37.88
A century ago,	8.22	As we all know	30.62

Table 4.15 Underused and overused phrases in a corpus of learners' English ranked by G score

There is a close correlation between the expressions overused by non–native speakers and the lists of expressions in which many students of English are drilled. The 'carrying over' of literal translations of phrases from the students' first language is a much less significant factor. Milton uses this data in the construction of a hypertext-based computer-assisted language learning (CALL) system.

Kilgarriff (1996a) having compared the chi–square and G–square measures, preferred the G–square. Dunning (1993) points out that most vocabulary items are rare, and thus words in the text are not normally distributed. The advantage of the G–square or log likelihood measure is that it does not assume the normal distribution.

3.2.16 Dispersion

Another set of useful measures for dealing with words and multi–word units is dispersion measures. These show how evenly or otherwise an item is distributed through a text or corpus (Lyne 1985, 1986). In early studies where word lists were compiled, the items were ranked by overall frequency or range (number of subsections in which the word appeared) alone. Range and simple frequency can be replaced by Juilland's D measure (Juilland et al. 1970), originally developed for Spanish texts, which takes into account not only the presence or absence of an item in each section of a corpus but also its subfrequency in each section. If the D measure for a word is multiplied by its frequency F, a value U called the 'usage coefficient' is obtained. Dispersion measures have also been produced by Carroll (1970) and Rosengren (1971), which are referred to as D_2 and S respectively. These measures also produce corresponding usage coefficients called U_m and KF. Juilland and Carroll derive D and D_2 first, then multiply by frequency to give U and U_m. Thus, the usage statistics can only be as accurate as their corresponding dispersion measures.

In order to evaluate the three dispersion measures, Lyne (1985) divided the French Business Correspondence (FBC) corpus into five equal sections. D is calculated for a given word. If the corpus consists of five equally large subsections, we call the subfrequencies of that word (frequency in each sub-section) x_1, x_2, x_3, x_4, and x_5. x_i is used to denote each of these subscripts in turn. Their mean is called $\bar{x}$. We then find the standard deviation s of the subfrequencies, using the formula

$$s = \sqrt{\frac{\sum (x_i - \bar{x})^2}{n}}$$

To evaluate this formula, we first find $\bar{x}$, the mean subfrequency of the word, by adding together all the values of x_1 (x_1, x_2 etc., each of the individual subfrequencies) then dividing by n, the number of subsections in the corpus. Then, for each value of x_i, we find the difference between $\bar{x}$ and that value of x_i, and square this difference. The sum of these squares (taking into account all values of x_i) is then divided by n, to yield the variance. The square root of the variance is the standard deviation. To account for the fact that s increases with frequency, s is divided by the mean subfrequency $\bar{x}$ to give a coefficient of variation, V:

$$V = s / \bar{x}$$

The coefficient of dispersion, D, is derived from V, and is designed to fall in the range 0 (most uneven distribution possible) to 1 (perfectly even distribution throughout the corpus), as follows, where n is the number of corpus subsections.

$$D = 1 - \frac{V}{\sqrt{n-1}}$$

Carroll's D_2 measure uses information theory. Entropy H is calculated as follows:

$$H = \log_2 F - \left(\sum_{i=1}^{n} x_i \log_2 x_i \right) / F$$

We first find the logarithm to the base 2 of the frequency of the word of interest. Then for each value of x_i we multiply by the logarithm to the base 2 of

x_i. This sum is first divided by the word frequency F, and the result of this is subtracted from the logarithm of F calculated earlier. H is then divided by $\log_2(n)$ to yield the 'relative entropy' or dispersion measure D_2.

$$D_2 = H / \log_2 n$$

To calculate Rosengren's S measure, we start by calculating the usage coefficient KF, and then divide KF by frequency F to obtain the dispersion measure S. KF is calculated using the formula

$$KF = \frac{1}{n}\left(\sum_{i=1}^{n} \sqrt{x_i}\right)^2$$

Thus, we calculate the square root of each subfrequency, sum them, square the result and finally divide by n (the number of subsections). The dispersion measure is always in the range 1 (perfectly even distribution) to $1/n$ (most uneven distribution possible, where all occurrences of a word are found in the same subsection).

Lyne (1985) performed an empirical investigation of the three measures. The word rankings for 30 words taken from the FBC corpus, all with a frequency of 10, were found for each of the three measures, and the correlation coefficients between each pair of measures were found. Pearson's product moment correlation coefficient was 0.97 for D vs D_2, 0.92 for D vs S, and 0.96 for D_2 vs S. Thus, all three measures were highly correlated. Lyne then calculated theoretical values for all three measures for a word with frequency of 10. There are 30 different ways 10 words can be distributed across five subsections (for example, all 10 words occur in one subsection, or two words occur in each subsection). For each of these 30 possible distributions, D, D_2 and S were calculated. D_2 gave consistently higher values (except for equal values at the end points of the comparison graph when the score was 0 or 1) but this difference is constant and had no effect on the rankings. D_2 penalises instances where one or more corpus subsections have no instances of the word (referred to as 'zeros'). When comparing S and D for each of the 30 possible distributions, S was consistently higher, reflecting the fact that its minimum value was 0.2 rather than 0. S penalises zeros even more than D_2. In fact, D_2 and S penalise not only zeros, but any subfrequency which is substantially below the mean. Lyne thus found that D was the most reliable of the three measures.

It is not possible to use these measures if the corpus is subdivided into sections of different sizes. Since no word count is ever based on a truly random sample of the target population the sample frequencies are generally overestimates of the true population frequencies, except that the highest-frequency items are underestimated as a consequence of the latter phenomenon (Hann 1973).

Chi-square can also serve as a measure of evenness of distribution. Equiprobable distributions are characterised by the same chi-square value. The general formula for chi-square is

$$chi - square = \sum \frac{(O - E)^2}{E}$$

Here the observed values (O) are the subfrequencies x_1 to x_5 (x_i) and the expected or estimated value (E) is the mean subfrequency $\bar{x}$. Thus

$$chi - square = \sum \frac{(x_i - \bar{x})}{\bar{x}}$$

3.2.17 Hayashi's quantification method

Hayashi's quantification method type III, a method of multivariate statistical analysis, has been applied by Nakamura and Sinclair (1995) to the study of collocations. They counted the statistically significant collocates of the word *woman* in four subsections of the Bank of English corpus, then used Hayashi's method to cluster the collocates of *woman* with one or other of the subsections of the corpus. The Bank of English is a corpus of current English compiled by Cobuild. In total, it consists of over 200 million words, but in this study just four subcorpora were considered, taken from books in a variety of fields, *The Times* newspaper, spontaneous speech and BBC World Service broadcasts. These four experimental subsets of the Bank of English comprised over 53 million tokens in total.

Texts and corpora are generally classified by factors external to the text itself such as the name of the author or the time of writing. Much less use is made of the patterns of the language of the texts themselves, which constitutes internal evidence, although the work of Biber (1988) has highlighted a number of linguistic features within texts that might be used to differentiate those texts. Hayashi's method also deals with internal evidence, determining the lexical structure of the corpus by quantifying the distribution across texts of linguistic features associated with a word or phrase of interest, in this case the collocates of *woman*. The word *woman* was chosen because it was a mid-frequency term, well distributed across the subcorpora. Since gender is a factor in linguistic variation, Nakamura and Sinclair felt that a study of *woman* was likely to produce variation between texts. Initially those collocates occurring within a window four words on either side of *woman* and which were significant with $p<0.001$ in at least one of the subcorpora were extracted, yielding a total of 282 such collocates. Alongside each collocate was stored its frequency in each subcorpus, this raw data being amended as follows:

1. If a collocate was not significant in a particular subcorpus, its frequency figure for that subcorpus was reduced to nil
2. the frequency figures were normalised to account for the subcorpora being of different sizes.

The amended frequency list became the input for Hayashi's method. The method simultaneously provided each subcorpus and each collocate with a set of numeric quantities. The number of quantities in a set is always one less than the number of subcorpora, giving three quantities for each data item, and enabling each data item to be plotted on a graph with three axes, each one corresponding to a member of each set of quantities. Subcorpora and collocates can be plotted on the same graph, and items with similar distributional properties appear close together. For example, the BBC subcorpus was clustered with adjectives dealing with country, citizenship or religion, such as *Palestinian* or *Korean*, verbs related to violence or crime such as *injure, suspect* or *convict*, and proper nouns. All these concepts are frequently in the news. The other subcorpora were clustered with very different sets of collocates, showing that Hayashi's method forms the potential basis of a powerful technique for the comparison of texts. For example, the book subcorpus was clustered with words specifically related to womanhood, such as *feminine, pregnancy* and *priestess*.

4 CONCORDANCING TOOLS USING MORE ADVANCED STATISTICS

Having covered a range of useful statistics, let us now look at some advanced concordance packages to see how widely implemented these measures are in retrieval software packages. The concordance packages which will be described in the following sections are WordSmith which uses the chi-square measure to identify 'keywords' (words whose frequency is unusually high in comparison to some norm, and thus help in genre identification), CobuildDirect which uses mutual information and the *t* score to identify significant collocates, Lexa which provides information on lexical densities (how a word is dispersed or clumped throughout the text), TACT, and the Hua Xia concordancer for Chinese text.

4.1 WordSmith

WordSmith[6] was developed by Scott (1996), and is designed for linguists and language teachers. It comprises three main tools called Wordlist, Concord and Keywords.

Wordlist is based on a corpus of about two million words called the MicroConcord corpus collections. It generates word lists in alphabetical and frequency order so the lexical content of different texts can be compared. In the alphabetical list, all the words in the corpus are arranged alphabetically, and the user can scroll through to read the frequency of each word in the corpus. The frequency list lists all the words in the corpus from the most frequent downwards, and for each word gives its rank, frequency in the corpus and the percentage of words in the corpus constituted by that word. For example, the entry for *the* is 1,141,967; 6.93 per cent. For a given text, Wordlist can provide the following statistics: the total number of words (tokens); the number of different words (types); the type-token ratio; the average word length in letters; the average sentence length in words and the standard deviation; the average paragraph length in words and the standard deviation; and the number of sentences.

Concord creates concordances, providing lists of the search word in context. It identifies the commonest phrases or clusters around the search word in the concordance, including the way they are normally punctuated and their frequency in the corpus. For example, the three-word clusters containing the search word *however* include *However it is* (frequency 22) and *there are, however* (frequency 16). Concord finds collocates of the search word: the most frequent words to its left and right. It can also display a graphical map showing where the search word occurs in the corpus and allowing the user to literally see if it is clumped or evenly distributed.

The keywords identified by the Keywords concordancing tool are words whose frequency is significantly higher in a small text of interest rather than in a larger general corpus. The identification of such keywords will help to characterise the subject matter or genre of the text. For example, one text concerned with architecture contained significantly more architectural terms than the general corpus, such as *design, drawing* and *esquisse*. The identified keywords are given in order of highest score first, where the user is offered a choice of scores between chi-square and Dunning's log likelihood. One entry might be *school 67 0.33% 466 0.02% 723.4 p. 0.01*, showing that the frequency of *school* in the small wordlist is 67, where it comprises 0.33 per cent of all word occurrences. Its frequency in the reference corpus is 466, comprising just 0.02 per cent of all word occurrences. The chi-square value is 723.4, which is significant at the 0.01 per cent level.

4.2 CobuildDirect

CobuildDirect[7] enables the user to choose between mutual information (MI) and the *t* score, and the output is the top 100 most significant collocates of a chosen keyword according to the selected score. Using MI, the word *software* was found to have the collocates OCR (optical character reader), *shrinkwrap*, and *antivirus*. Using the *t* score, the top collocates were *computer* and *hardware*. The *t*-score collocates tended to have a higher frequency than the MI collocates. The CobuildDirect corpus sampler query syntax allows one to specify word combinations, wildcards, part-of-speech tags and so on. A query is made up of one or more terms concatenated with a + symbol, where, for example, *hell + hole* will search for the word *hell* immediately followed by the word *hole*. Other available retrieval options include for example *dog +4 bark*, which will retrieve all occurrences of *dog* occurring not more than four words away from *bark*. *blew@* will retrieve all grammatical variants of *blew*, such as *blow, blowing* and *blew* itself. A trailing asterisk will cause truncation to be performed, so that *cut** will retrieve references to *cut, cuts* or *cute*, which all commence with the same three characters. The appropriate grammatical tag of a retrieval word can be specified, using the syntax *word/TAG*, for example, *cat/NOUN*.

4.3 ICAME's Lexa

Lexa[8] was devised by Hickey (1993) as a lexical and information retrieval tool used for corpora, particularly those containing historical texts. Its lexical analyses include tagging, lemmatisation, statistics (including number of characters, words or sentences, and lexical densities, all of which can be saved to a file), and it can also generate concordances (Kirk 1994). For information retrieval, Lexa can perform pattern-matching for user-specified strings in files.

4.4 TACT

The TACT concordancer[9] enables the production of KWAL (Key Word and Line) and KWIC (Key Word in Context) concordances, word frequency lists, and graphs of the distribution of words throughout the corpus. TACT can produce collocation lists, identifying statistically significant collocations using Berry-Rogghe's z score as described in Section 3.2.1. Groups of words specified by the user, such as those belonging to a common semantic category, can be searched simultaneously.

4.5 The Hua Xia concordancer for Chinese text

The Hua Xia corpus of Chinese text consists of 2 million characters, and is composed of nearly 200 weekly issues of *Hua Xia Wen Zhai* (*China News Digest Computer Magazine*), an electronic magazine published by the Chinese overseas community since 1991. A concordancer was created to retrieve sequences of Chinese characters from this corpus, where the output may be in order of appearance in the text, or sorted according to the alphabetical order of the romanised (Pinyin) equivalents on either the right or the left of the retrieved sequence. The length of the context on each side for each retrieved sequence may be varied. The preliminary input to this concordancer is carried out using a Chinese word processing package called NJstar (1991–2), which allows 16 different modes of producing Chinese characters from a standard keyboard. The output of the Hua Xia concordancer must also be displayed using NJstar for examination.

The mutual information was found for sequences of two characters, and the simple frequency counts were found for sequences of two to four characters within a sample of 1.4 million characters. The use of these measures facilitated the identification of Chinese four-character idioms and proper nouns of two to four characters. The resulting lists of proper nouns were incorporated into the concordancer to produce a news retrieval tool (Oakes and Xu 1994).

5 SUMMARY

Concordancing, collocations and dictionaries are closely-related concepts. A concordance is a list of words extracted from an underlying corpus, each displayed with their surrounding context. Collocations are pairs or groups of lexical items which occur in the same context more frequently than would be expected by chance. Statistics allow collocations to be quantified according to the strength of the bond or bonds linking the collocating words, while concordances enable the visual display of collocations. A wide variety of different statistical measures for the identification of collocations in a corpus have been described in this chapter, which differ according to the size of the collocational sets they uncover and whether they are used with monolingual or bilingual parallel-aligned corpora. In this chapter, the correspondence of words in parallel corpora which are putative translations of each other has been regarded as a form of collocation across languages. Many statistical scores are based on the values found in the contingency table. Collocations can be also be identified syntactically, since certain syntactic patterns have been found to occur regularly in collocations. Two important categories of collocations are idiomatic collocations, which can be collated into dictionaries for learners of English as a second language, and technical terms which can be collated into domain-specific glossaries. A number of concordancing tools now make use of statistics such as mutual information or the t score for the display of collocations, or scores related to the chi-square measure to build word lists of genre-specific terms.

6 EXERCISES

1. The Kulczinsky coefficient (KUC) between various word pairs was found to be as follows:

English word	French word	Kulczinsky coefficient
mechanical	mécanique	0.99
mechanical	mécanisme	0.98
engine	mécanique	0.96
driver	mécanicien	0.91
engine	mécanicien	0.87
engine	machine	0.75
mechanical	machine	0.64
trouble	ennui	0.59
engine	ennui	0.48
driver	locomotive	0.32

Using this information and the best match criterion of Gaussier, Langé and Meunier, infer the likeliest translation of *engine*.

2. The following table shows a small parallel-aligned corpus.

Le chat blanc.	The white cat.
Le chien noir.	The black dog.
Le chat grand.	The big cat.
Le chien grand.	The big dog.

a) Using this data, create contingency tables to find the cubic association ratio (MI3) between *chat* and each of the following: *black, big, cat, dog, the* and *white.* What is the most probable translation of *chat*? The higher the MI3 value, the more likely that the two terms are translations of each other.

$$\log_2(0) = -\infty, \log_2(1) = 0, \log_2(2) = 1, \log_2(4) = 2, \text{ and } \log_2(8) = 3.$$

b) Repeat the experiment using mutual information (MI). In what way is the result surprising?

3. The following table shows the number of times five overlapping word sequences were found in a corpus:

Sequence	Occurrences
a	27,136
a number	102
a number of	51
a number of times	20
a number of times we	8

Using this data and the cost criterion of Kita et al., calculate which of the sequences should most properly be considered as a collocation.

7 FURTHER READING

The best general introduction to word association measures and their evaluation is given by Daille (1995). The paper by Kilgarriff (1996a) can be seen as complementary to this.

NOTES

1. Information about resources for collocation researchers can be found on the internet, accessed via Jennifer Lai's homepage at http://www.ed.uiuc.edu/students/jc-lai/fall95/collocations.html
2. An MS-DOS version of the OCP called Micro-OCP became available in 1989. It can be bought from Oxford University Press, Walton Street, Oxford OX2 6DP, UK.
3. WordCruncher can be bought from: Electronic Text Corporation, 780 South 400 East, Orem, Utah 84058, USA.
4. Assuming a simple *word_tag* style of annotation.
5. For further information about the availability of XTRACT, contact Eric Siegel at evs@cs.columbia.edu
6. WordSmith can be ordered through the internet at http://www1.oup.co.uk/elt/software/wsmith
7. CobuildDirect can be ordered through the internet at http://titania.cobuild.collins.co.uk/direct_info.html
8. ICAME's Lexa can be ordered through the internet at http://www.hd.uib.no/lexa-ftp.html
9. TACT can be obtained through the Contact Centre for Computing in the Humanities at the University of Toronto (email cch@epas.utoronto.ca) or ICAME (email nora.hd.uib.no)

Literary detective work

1 INTRODUCTION

This study will end with a detailed examination of literary detective work. While not strictly corpus linguistics, this work shares many affinities with corpus linguistics, particularly its focus on manipulating large data sets. Consequently, an examination of literary detective work can help us see how a wide range of the statistics covered in this book can be applied to 'real' language. A survey of literary detective work will show us what studies we might make with a corpus.

One aspect of literary detective work on the computer concerns the statistical analysis of writing style. In order to perform a computer analysis of style, most studies concentrate on features which are easily identifiable and can be counted by machine. The features most frequently used are:

- word and sentence length
- the positions of words within sentences
- vocabulary studies, based on what Hockey (1980) calls the choice and frequency of words
- syntactic analysis, including the additional tasks of defining and classifying the syntactic features.

Morton's (1978) criteria for the computation of statistical measures of style are given in Section 2.1.

Kenny (1982) lists a number of examples of applications in which statistics are used to compare and contrast the stylistic features of various texts. Foremost among them are studies of authorship attribution, covered in this chapter in Sections 2.2 to 2.10, using such techniques as examining the position and immediate context of word occurrences. For example, Morton (1965), as described in Section 2.4, measured the number of times *and* occurs at the start of a sentence. Related studies have helped determine the chronology of texts, such as Zylinski's (1906) dating of Euripides' plays by the increasing number of unresolved feet in his verses.

Statistical methods can be used not only to distinguish between different authors, but also to look at variations within the style of a single author. An example of this is that a single author, such as Kierkegaard, might write using different pseudonyms and adopt noticeably different writing styles for each pseudonym, as described in Section 2.13 (McKinnon and Webster 1971). Bailey (1969) states that chronology problems are also closely related to the problem of disputed authorship. Described in Sections 2.11 to 2.12, these are concerned with specifying the order of composition of the works of a particular author, where critics speak of the 'early' or 'late' styles of that author. As well as determining the position of documents in time, statistical studies can be used to estimate the place of origin of texts. Leighton (1971) calls this forming a literary geography of texts, and some of his work will be described in Section 2.15. Similar techniques are also used in forensic stylometry (Morton 1978), which includes determining the authorship of anonymous letters in serious crimes or distinguishing between real and fabricated confessions. Case studies will be provided in Section 2.16. One study which stands alone in its use of a syntactically annotated corpus for author discrimination will be described in Section 2.17.

Another aspect of literary detective work is determining which languages are related to each other, both in cases where the vocabulary of their common ancestor language is known and in cases where the vocabulary of the antecedent language must be inferred. The key to determining the degree of the relationship between languages is the identification of cognate word pairs, where a word in one language is lexically related to the word in another language and has the same meaning. A knowledge of cognates and their rates of disappearance from languages can help show the length of time elapsed since two related languages diverged, using a model related to the process of radioactive decay (Dyen, James and Cole 1967) and described in Section 3.2.

The decipherment of Linear B will be discussed in Section 4.1 as an example of translation without the aid of a computer, to illustrate the type of processes which might be facilitated by use of an electronically stored corpus, such as the identification of unknown prefixes and suffixes by the comparison of words which are similar but not identical. Even with the availability of computer corpora, there is a need for human intuition and background knowledge as well as the processing power of the computer. This will be illustrated in Section 4.2, a description of the decipherment of runes on the cross at Hackness (Sermon 1996). Finally, the translation of English into French by purely computational means (Brown et al. 1990) will be covered in Section 4.3.

2 THE STATISTICAL ANALYSIS OF WRITING STYLE
2.1 Morton's criteria for the computation of style
Morton (1978) describes how the largest group of words in the vocabulary of a text is the words which occur infrequently, being used perhaps only once (the

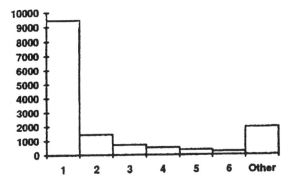

Figure 5.1 Numbers of words at various frequencies of occurrence in a 100,000-word sample of the British National corpus

hapax legomena). Although Morton does not make the distinction, there are words which are intrinsically rare because they are obscure, out-of-date or specialist terms, and those which are unlikely to occur just because of the frequency features of open-class words. Examples of each type of rare word would be *fustigate* and *trolley* respectively. Figure 5.1 is a plot of the number of times a word might occur in a 100,000-word sample of written text from the British National corpus (horizontal axis) against the number of words which have this frequency of occurrence (vertical axis). The frequency of occurrence of the words grouped under 'other' ranges from 7 to 7447. The figure shows that the overwhelming majority of words in a corpus have a low frequency of occurrence.

These words that are hapax legomena tend to be interesting, reflecting such factors as the background, experience and powers of mind of an author, and conveying delicate shades of meaning. Many qualitative studies of literature have focused their attention on rare words, since the commonest words tend to be connectives, pronouns and articles, words whose meaning and use are much more prescribed and conventional than is the case with rare words. However, in statistical studies, the examination of occurrences of rare words tends to be less fruitful than the study of common words, since their low rate of occurrence makes them difficult to handle statistically. Any stylistic trait which is to be used in the determination of authorship must be sufficiently frequent. To test any hypothesis statistically, using, for example, the chi-square test, we need a minimum of five occurrences, and ideally at least ten. Similarly, any stylistic trait which is related to the use of nouns is unlikely to make a good indicator of authorship, since the use of nouns is more closely tied to subject matter than to an individual author.

For a trait to be useful in the determination of authorship, it must be one that can be numerically expressed. For example, some New Testament writers show Semitic influences, but when considering the number of Semitisms in a given book of the New Testament as estimated by different scholars, the differences in classification are so large that no consistent count is possible. Thus, the occurrence of Semitisms cannot be numerically expressed exactly enough to be used in the determination of authorship by quantitative methods.

In stylometric studies the placing of words can be described in two ways. A word can have a positional reference, where we say it occurs, for example, as the first, second or last word in a sentence. Alternatively, a word can have a contextual reference, where we say, for example, that the word occurrence is preceded by word X or is followed by word Y. Word order can vary much more in inflected languages such as Greek, where it can provide the basis for studies of word mobility. Another approach, taken in relatively uninflected languages such as English, is the study of immediate context, particularly the study of collocations formed by the successive occurrence of two frequent words.

In summary, the total vocabulary used by an author, which depends heavily on the use of rare words, is not an effective means of discriminating between writers' styles. The selection and frequency of common words is much more effective, especially when there are large amounts of text to sample and few contenders for its authorship. Morton concludes that by far the most effective discriminator between one writer and another is the placing of words. The absolute position of words within a sentence is the most effective method for inflected languages, and the immediate context of words is most effective in uninflected languages. The only disadvantage of positional studies is that they are dependent on punctuation, which may not always be given in early texts.

Certain word pairings such as *and/also*, *since/because* and *scarcely/hardly* arise from the fact that the storage mechanism is the human brain, where they may be stored in similar locations and retrieved in response to similar cues. The first reference to such word pairs, now called **proportional pairs**, was by Ellegård (1962). Such pairs are useful in comparing literary texts, since different authors tend to use one member of a pair at a higher rate than the other. In terms of effectiveness, proportional pairs fall between the use of frequent words and the word placement techniques.

2.2 Early work on the statistical study of literary style

In 1851 a suggestion was made in a letter by de Morgan that the authenticity of some of the letters of St Paul might be proved or disproved by a comparison of word lengths. This is the oldest surviving reference to the creation of a scientific stylometry. Morton (1978) reports that de Morgan's letter contains many of the basic principles of stylometry, the use and description of samples, the disregarding of the meaning of words and the concentration of their occurrences.

In 1887 T. C. Mendenhall studied word length to compare the works of Shakespeare, Bacon, Jonson, Marlowe, Atkinson and Mill. He constructed the word spectrum (frequency distribution of word length) for each of these authors and showed that texts with the same average word length might have different spectra, the frequency spectrum being a graph of number of occurrences against word length. For example, for texts written by J. S. Mill, the mode was two characters, and in *Oliver Twist* the mode was three characters. However, the average word length for texts by J. S. Mill was 4.775 characters, compared with texts by Charles Dickens where the average word length was 4.312 characters. Later, in 1901, Mendenhall showed that Shakespeare had a very consistent frequency spectrum both in poetry and prose, distinct from that of Bacon, but not possible to distinguish from that of Marlowe (Kenny 1982).

In 1867 Lewis Campbell produced a battery of stylistic tests designed to differentiate between works of Plato written at different dates. He used word order, rhythm and lists of words occurring just once in the text. Campbell's ideas were confirmed by Ritter in 1888. Lutoslawski, also working with the texts of Plato, listed 500 potential stylometric indicators of date such as 'interrogations by means of *ara* between 15 per cent and 24 per cent of all interrogations'. Sherman, in 1888, rather than dating works by a single author, suggested that related methods could be used to observe the evolution of language as a whole. He found both shorter sentences and a greater proportion of simple to compound sentences in more modern works.

Between 1887, when Mendenhall did his work, and the present day, the science of statistics has flourished. Techniques have been developed to look for significant differences in data sets as opposed to random fluctuations, such as the *t* test described in Chapter 1, Section 3.2.1. The notion of sampling has been developed, to enable generalisations to be made without examining the whole population. Williams (1970) estimated that Mendenhall's results could have been obtained using a sample one-tenth of the size actually used. The advent of computers since the Second World War has facilitated calculations and the provision of data such as word lengths, and the handling of the large quantities of texts used in modern corpus linguistics.

2.3 The authorship of *The Imitation of Christ*

Williams (1970) reports that Yule (1939), working without the aid of a computer, used sentence length to decide that the 15th-century Latin work *De Imitatione Christi* was probably written by Thomas à Kempis rather than Jean

	De Imitatione Christi	**Kempis**	**Gerson(A)**	**Gerson(B)**
Mean	16.2	17.9	23.5	23.4
Median	13.8	15.1	19.4	19.9
Sentences over 50 words	15	22	66	68

Table 5.1 Sentence lengths in *De Imitatione Christi* and works known to be by Kempis and Gerson

Charlier de Gerson. He compared four samples, each of about 1200 words, one taken from *De Imitatione Christi*, one taken from works known to be by Kempis and two taken from works known to be by Gerson. His results are summarised in Table 5.1. The data for the *De Imitatione Christi* sample is much closer to that of the Kempis sample than that of the Gerson sample.

In 1944, Yule used a study of vocabulary to tackle this same problem of disputed authorship. He produced and used a measure of 'vocabulary richness' called the **K characteristic**. Yule's K characteristic is a measure of the probability that any randomly selected pair of words will be identical. To avoid very small numbers, Yule multiplied this probability by 10,000. His actual formula is

$$K = 10,000 \times (M2 - M1)/(M1 \times M1)$$

where $M1$ and $M2$ are called the first and second moments of the distribution. $M1$ is the total number of usages (words including repetitions), while $M2$ is the sum of all vocabulary words in each frequency group, from one to the maximum word frequency, multiplied by the square of the frequency. $M0$, not used in the formula, is the total vocabulary used, not including repetitions. For example, imagine a situation with a text consisting of 12 words, where two of the words occur once, two occur twice and two occur three times. $M0$ is six, $M1$ is 12, and

$$M2 = (2 \times 1^2) + (2 \times 2^2) + (2 \times 3^2) = 28$$

Yule's characteristic increases as the diversity of vocabulary decreases, and thus it is more a measure of uniformity of vocabulary than of diversity. Williams (1970) suggests that Yule's formula should be modified by taking the reciprocal and omitting the (arbitrary) multiplication factor to produce 'Yule's index of diversity', $(M1 \times M1)/(M2 - M1)$.

Yule (1944) studied the nouns in the three samples of text taken from *De Imitatione Christi*, Kempis and Gerson according to the following criteria: total vocabulary size, frequency distribution of the different words, Yule's characteristic, the mean frequency of the words in the sample and the number of nouns unique to a particular sample. The frequency distributions for the three samples were very similar for frequently used words, but the number of rarely used words was greatest for Gerson, then Kempis, then *De Imitatione Christi*. Yule's other data is summarised in Table 5.2. In each case the samples for

	De Imitatione Christi	Kempis	Gerson
Total vocabulary	1168	1406	1754
Unique nouns	198	340	683
Mean frequency of words	7.04	5.83	4.67
Yule's characteristic	84.2	59.7	35.9

Table 5.2 Total vocabulary, mean frequency of words and Yule's characteristic for three samples of text

Kempis and *De Imitatione Christi* resembled each other more closely than Gerson and *De Imitatione Christi*.

Yule made three contingency tables, the concept of which has been described in Chapter 1, Section 4.1. In Yule's tables a value of 37 in the box with coordinates *De Imitatione Christi* 3–5, Kempis 0, showed, for example, that of the 2454 nouns occurring in at least one of the three samples, 37 occurred between three and five times in *De Imitatione Christi* but did not occur at all in Kempis. The three tables were *De Imitatione Christi* x Kempis, *De Imitatione Christi* x Gerson and Kempis x Gerson, where the author occurring before the x was compared with the author coming after. Using these contingency tables, Yule calculated Pearson's coefficient of mean square contingency, (see Agresti 1990) where a coefficient c can only take the value zero for completely independent samples. The results are shown in Table 5.3, where the highest value of c is for the table of *De Imitatione Christi* x Kempis.

De Imitatione Christi x Kempis	$c = 0.71$
De Imitatione Christi x Gerson	$c = 0.61$
Kempis x Gerson	$c = 0.66$

Table 5.3 Pearson's coefficient of mean square contingency for three pairs of literary samples

Other data provided by Yule is reproduced in Table 5.4. In every case, Kempis appears to be the more likely author of *De Imitatione Christi*.

Quantity	Imitatio	Kempis	Gerson
Sum *M0*, Vocabulary	1168	1406	1754
Sum *M1*, Occurrences	8225	8203	8196
Sum *M2*, (ΣX^2).	577665	409619	248984
Mean *M* (*M1/M0*)	7.042	5.834	4.673
1000 3 *M0/M1*	142	171	214
Percentage of once–nouns	44.5	44.2	45.8
Characteristic K	84.2	59.7	35.9

Table 5.4 The sums and sundry data for a comparison of *De Imitatione Christi* and samples of Kempis and Gerson

2.4 The authorship of the New Testament Epistles

Morton (1978) reports that the Pauline corpus covers all 14 letters from Romans to Hebrews. He tackled two main problems, namely to establish the authorship of the individual epistles and to establish the integrity of the individual epistles. Some of them, such as Romans and II Corinthians, are regarded as being written by St Paul but with sections of the text open to question.

Comparing the sentence lengths in the Pauline epistles and in texts by other

Greek-language authors, he found that the variability in sentence lengths between the various epistles was greater than that found in any corpus by a single author. He suggested that Romans I and II, Corinthians and Galatians could be attributed to Paul, while the other Epistles were the work of up to six different authors. Kenny (1982) reports that a weakness in Morton's argument is that his methods also reveal anomalies in II Corinthians and Romans itself.

Morton (1965) studied the relative frequency of the word *kai* (the commonest word, meaning *and*) in the Greek epistles of the New Testament generally attributed to St Paul. Table 5.5 shows Morton's results for 11 of the epistles, according to the use of *kai* as a percentage of the total number of words, and the percentage of sentences containing *kai*. The data appears to fall into two groups, with Romans, Galatians and Corinthians I and II producing similar results as if written by one author, and the other epistles producing significantly different results as if written by another author or authors.

	kai as a percentage of all words	Percentage of sentences containing *kai*
Romans	3.86	33.7
I Corinthians	4.12	41.6
II Corinthians	4.41	40.7
Galatians	3.22	29.3
Ephesians	5.67	67.0
Philippians	6.56	51.7
Colossians	6.27	51.8
Thessalonians I	6.73	58.0
Timothy I	5.72	54.2
Timothy II	5.41	48.9
Hebrews	5.16	49.1

Table 5.5 The usage of *kai* in epistles usually attributed to St Paul

These studies of sentence length and Greek function words such as *kai* showed, according to Morton, that St Paul wrote only four of the New Testament epistles traditionally attributed to him, since the others show less consistency of style. However, when Ellison (1965) used these same attributes on English texts, they showed that James Joyce's *Ulysses* and Morton's own essays must each have been written by several authors. Other criticisms of Morton's methods were that he encoded his Greek texts with modern punctuation and also that sentence length varies considerably within random samples of text. Morton later suggested new stylistic criteria such as the part of speech of words coming last in a Greek sentence, the number of words between successive occurrences of *kai*, the relative occurrence of genitive and non-genitive forms of the Greek pronoun *autos* (*he*), and pairs of words in English which collocate such as *of course* and *as if* (Oakman 1980).

2.5 Ellegård's work on the Junius letters

The Junius letters are a series of political pamphlets written between 1769 and 1772 under the pseudonym Junius. The authorship of the letters has been attributed at various times to no fewer than 40 different people. Ellegård (1962) compared the anonymous Junius letters and their 157,000-word context with 231,000 words of text by Sir Philip Francis (the most probable contender for authorship of the Junius letters) and other 18th-century English writers. He found that Yule's sentence-length test was not sensitive enough to discriminate between many authors. The K characteristic could not be used since it requires large text samples of at least 10,000 words and some Junius samples are less than 2000 words.

Ellegård presented a list of 458 words and expressions (mainly content words) designated either 'plus' (occurring more frequently in the Junius letters than in work by other authors) or 'minus' (occurring less frequently in the Junius letters than in work by other authors) indicators of style. He calculated their occurrence in all the sample texts. The ratio of usage of a word in samples of an author's text compared with its usage in a representative sample of contemporary texts is called the **distinctiveness ratio**. The plus words have a distinctiveness ratio above 1, while the minus words have a ratio below 1. For example, the relative frequency of the word *uniform* (when used as an adjective) was 0.000280 in the sample of Junius texts, but only 0.000065 in the comparison sample of one million words. The distinctiveness ratio of *uniform* is then 0.000280/0.000065 which is about 4.3, showing that *uniform* is a plus word. In order to produce a reliable 'testing list' of plus and minus words, a text of at least 100,000 words is required. However, once this list has been produced, it can be used to test much smaller samples. Ellegård's study of 'plus' and 'minus' words showed that Sir Philip Francis was the most likely author of the Junius letters.

As Ellegård did not use a computer for counting the words, he had to rely on his own intuition as to which ones occurred more or less frequently than expected. The study would have been more comprehensive if he had used a computer for the counting and to have counted all words, not just those he specifically noticed (Hockey 1980).

Ellegård also identified about 50 pairs or triplets of approximate synonyms such as *on/upon*, *kind/sort*, *and/also*, *since/because* and *scarcely/hardly*. Their patterns of usage in the Junius letters showed strong correspondence with the textual characteristics favoured by the author Sir Philip Francis, while the other authors were clearly distinct.

Austin (1966) also found lists of 'plus' words which were able to discriminate between the work of Robert Greene and Henry Chettle, both possible authors of *The Groatsworth of Wit* written in 1592. Once all spelling and inflectional variants of each word had been merged, 'plus' words were taken to be those which fulfilled the following criteria:

1. the word must occur at least ten times in one author
2. the frequency per 1000 words in one author must exceed the corresponding frequency in the other by 1.5 – the differential ratio
3. the ratio of variation within the body of works of one of the authors must be lower than the differential ratio.

Fifty words were found which fulfilled all three criteria. Austin's work demonstrates the three stages of automated authorship studies. Firstly, the computer can be used to look in large bodies of text for some stylistic feature with discriminating power; secondly, the disputed texts are searched for the presence or absence of this feature; and finally, a statistical analysis is performed to test the validity of the findings (Oakman 1980).

2.6 The Federalist papers

The so-called Federalist papers were published in newspapers in 1787–8 to persuade the population of New York state to ratify the new American Constitution. Published under the pseudonym Publius, their three authors were James Madison, John Jay and Alexander Hamilton. In 1804, two days before his death in a duel, Hamilton produced a list of which essays were written by which author. He left this list in the home of a friend called Egbert Benson. In 1818 Madison claimed that some of the essays in the Benson list attributed to Hamilton were in fact written by himself. Altogether, it was agreed by both Hamilton and Madison as well as later historians that Jay wrote five of the essays, 43 were written by Hamilton and 14 by Madison. Another 12 were disputed, and three were written jointly by Hamilton and Madison (Francis 1966). In Sections 2.6.1 and 2.6.2 we will look at two contrasting approaches for determining the authorship of the disputed papers.

2.6.1 The Bayesian approach of Mosteller and Wallace

The stylistic features of the disputed papers could be compared with the features in the papers of known authorship. Unlike the case of the Junius letters which might potentially have been written by any of a large number of authors, in the case of the disputed authorship of the Federalist papers, there are only two candidates, Hamilton and Madison. Mosteller and Wallace (1963) used Bayesian statistics, described in Chapter 1, Section 6, to determine which of the disputed papers were written by which author, hoping that these methods would lead to the solution of other authorship problems. Like Ellegård, Mosteller and Wallace combined historical evidence with statistical computation.

The problem of deciding which variables to use to discriminate between the writing styles of Hamilton and Madison is difficult, because both adopted a writing style known as **Addisonian** which was very popular in their day. Since the two authors may consciously have tried to imitate each other's style, stylistic features which can be easily imitated – sentence length, for example – had to be discounted from the analysis. In this respect, variations in the use of high-frequency words might be a reliable criterion, since authors probably would not be conscious of using these. It is also desirable that variables should not vary with context (for example, sentence length in legal documents is greater than in most other texts), but must have consistent rates over a variety of topics. The occurrence of function words was examined since these tend to be non-contextual, except for personal pronouns and auxiliary verbs. The styles of Madison and Hamilton were also found to vary significantly in the frequency of use of various function words, such as *by* (more characteristic of Hamilton) and *to* (more characteristic of Madison). Some low-frequency words were also used since they tended to occur almost exclusively in the work of one author. For example, in a sample of 18 Hamilton and 14 Madison papers, *enough* appeared in 14 Hamilton papers and no Madison papers, and *whilst* appeared in no Hamilton but 13 Madison papers. Altogether, 28 discriminating terms were found.

Two mathematical models exist which describe the frequency distribution of individual words across equal-sized portions of text. In the Poisson model, the occurrence of a given word is independent of the previous occurrence of that word – it depends only on its overall rate of occurrence. However, a writer might consciously avoid repetition of words on one hand, and on the other may repeat a word for emphasis or parallelism (Francis 1966). The tendency for the same word to appear in clusters is taken into account by the alternative negative binomial model (see Agresti 1990). The Poisson distribution is described by a formula which, if given the average rate of occurrence of a word in a text of certain length in words (a single parameter), will give the proportions of text sections of that length which have none, one, two, etc., occurrences of the word. The Poisson formula takes the form

$$p_n = \frac{\lambda^n \times e^{-\lambda}}{n!}$$

where e is the constant, and λ is the average number of times the word occurs per section of text. p_n is the proportion of text sections which have n occurrences of the word. $n!$ means n multiplied by $(n-1)$, then multiplied by $(n-2)$ and so on until we reach 1. Thus $3! = 3 \times 2 \times 1$. By convention, $0! = 1$. e to the power x is often called the exponential (exp) of x, and may be calculated using a typical scientific calculator. For example, if the average occurrence of a given word per section of text $(\lambda) = 0.1$, then the proportion of text sections with no occurrences at all of the word (p_0) is $e^{-0.1} = 0.905$, the proportion of text sections with just one occurrence (p_1) is $0.1 \times e^{-0.1} = 0.0905$,

$$p_2 = \frac{(0.1)^2 \times e^{-0.1}}{2} = 0.0045, \text{ and}$$

$$p_3 = \frac{(0.1)^3 \times e^{-0.1}}{6} = 0.0001.$$

The negative binomial formula requires two parameters: the average rate of occurrence of a word and its tendency to cluster (non-Poissonness parameter). The negative binomial distribution fitted the observed data better than the Poisson distribution.

Francis (1966) gives the following example of the use of Bayesian statistics. Imagine the word *also* is used 0.25 times per 1000 words by Hamilton and 0.5 times per 1000 words by Madison, and that both authors follow the Poisson model. Imagine too that the word *also* occurs four times in a 2000–word paper. The Poisson probabilities for a word from a 2000-word paper, Hamilton's rate per 1000 words being 0.25 and Madison's 0.5, for four occurrences are 0.00158 for Hamilton and 0.0153 for Madison. The calculation for Hamilton was performed as follows: since Hamilton used the word *also* 0.25 times per 1000 words, this means that he used *also* at a rate of 0.5 times per 2000 words (the length of the text of unknown authorship). This rate of 0.5 becomes λ in the Poisson equation. Since we are interested in the probability of four occurrences of *also* in a 2000-word text, n is 4, and the bottom line of the equation is 4! = 4 × 3 × 2 × 1 = 24. This gives

$$p_4 = \frac{(0.5)^4 \times e^{-0.5}}{24} = \frac{0.0625 \times 0.6065}{24} = 0.00158$$

Thus, it is more likely that Madison wrote the paper, with a likelihood ratio of 0.0153/0.00158 giving odds of about 10 to 1. If we had had a prior opinion that Madison was the more likely author with odds of 3 to 1, this new evidence would be multiplied by the prior belief to yield posterior odds of 30 to 1. The next piece of evidence that we will examine is that the word *an* appears seven times in the disputed paper. Using tables of the Poisson distribution we find that the likelihood ratio for this eventuality is 0.0437/0.111 which is about 3/8. The old posterior odds become the new prior odds, and these are multiplied by this new likelihood ratio to give new posterior odds of 80 to 1. This process continues by considering the usage of a range of words whose discriminating power is high.

One of the assumptions made in this simplified analysis given by Francis is that the average rates of usage of each word by each author (the parameters of the model) were known. However, the true rates are unknown. We can only count the rates of occurrence in each word in **available texts** known to be written by each author, but we cannot possibly obtain all texts ever written by each author, and some existing texts we would exclude as their authorship remains controversial. The parameters must be estimated from prior information (for example, existing studies of word rates) and sample data (94,000 words of text by Hamilton and 114,000 words written by Madison).

Since the exact values of the authors' word-usage rates are not known, Francis maintains that it is better to express the knowledge that we do have in the form of probability distributions rather than point estimates such as 0.25 and 0.5. Thus, rather than stating that Hamilton's rate for a given word is 0.25, we could state, using hypothetical figures, that there is a probability of 1/20 that his rate is less than 0.15, a probability of 1/10 that it is between 0.15 and 0.20 and so on. Once again, Bayes's theorem is employed to combine the information contained in the prior word-frequency distribution with the information contained in the samples of known works of the two authors about relative rates of word usage summarised as a likelihood ratio. Because of the imprecision in the prior information, Mosteller and Wallace chose several possible distributions to be sure that at least one fitted the true prior distribution accurately, and carried out their analysis using each of these in turn.

Thirty words were eventually chosen for the main study, since a pilot study showed them to be good discriminators between Madison and Hamilton. Some words were discarded because they had different rates of usage when used by Madison within and outside the Federalist corpus. The final list is given in Table 5.6.

Group B3A	upon
Group B3B	also an by of on there this to
Group B3G	although both enough while whilst always though
Group B3E	commonly consequently considerable(ly) according apt
Group B3Z	direction innovation(s) language vigor(ous) kind matter(s) particularly probability work(s)

Table 5.6 Final words and word groups used by Mosteller and Wallace

The results were reported not in odds but in log odds, which are the natural logarithms of the odds, to restrict the range of results to more manageable values. (Natural logarithms or $\log_e$ (sometimes written 'ln') can also be calculated using a typical scientific calculator.) Positive values indicate a verdict in favour of Hamilton, while negative values indicate a verdict in favour of Madison. The method of Mosteller and Wallace was checked by applying it to 11 papers known to be written by each author (eight taken from the Federalist source and three taken from external sources). The resulting data, when assuming a negative binomial distribution, reveals that every Hamilton paper has positive log odds and every Madison paper negative log odds. This is evidence that their method is accurate. When the method was employed to obtain log odds for the three joint and 12 disputed papers, for every prior distribution tested the resulting log odds were greatly in favour of Madison being their author.

Oakman (1980) notes that Mosteller and Wallace's expectation that the distribution of function words may be a general signifier of authorship has been thrown into doubt by Damerau (1975), who found a great disparity in the use of function words in samples of *Vanity Fair* and three American novels. Thus a particular word list worked well for the Federalist papers, but the same list cannot necessarily be used for other authorship studies.

2.6.2 Biological analogy and the Federalist papers: recent developments

Tweedie, Singh and Holmes (1994) show that statistical methods of authorship attribution can be used in conjunction with a neural network to provide an effective classification tool. Previous studies of the Federalist papers had used function words, and in their preliminary experiments Tweedie, Singh and Holmes also chose common words on the basis of their ability to discriminate between the work of Hamilton and Madison. These words were *an, any, can, do, every, from, his, may, on, there* and *upon*. The number of occurrences per thousand words of each of these words was found using the Oxford Concordance Program, described in Chapter 4, Section 2.4.2. These values were converted to *z* scores as described in Chapter 1, Section 2.4, so that each word had a rate that was normally distributed with a mean of 0 and a variance of 1. This ensures that each word contributes equally to the neural network training process.

Neural networks simulate the way neurons interact in the human body. Each neuron has a body, and receives stimuli from other neurons or the environment. The human neuron is either at rest or firing. It fires whenever the sum of the stimuli it receives exceeds a certain threshold, and when it fires, other neurons are stimulated or an action such as a movement is taken. Tweedie, Singh and Holmes used an array of computer-simulated neurons called a multilayer perceptron, which consisted of a so-called input layer of 11 neurons, each of which was stimulated to a degree which depended on the occurrence of one of the 11 marker words in a given text. These neurons were connected to a middle or 'hidden' layer of three neurons, and these in turn were connected to an 'output' layer of two neurons. One of these would fire if the network thought the text was written by Hamilton, and the other would fire if the network thought the text was written by Madison. The number of neurons in the input and output layer are clearly related to the number of possibilities within the task at hand, but there are no hard and fast rules for deciding on the most effective number of neurons in the hidden layer. A small part of this neural network is given in Figure 5.2, where the two input neurons which respond to the occurrence of *there* and *on* are shown, connected to one of the hidden layer neurons which in turn stimulates the two outer layer neurons.

These artificial neurons differ from human neurons in at least three important respects:

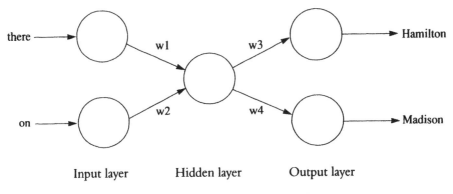

Figure 5.2 A section of the Federalist neural network

1. In the artificial network, the degree of stimulation afforded by one neuron to another to which it is connected depends on a weight which is continually updated during the training phase.
2. The weights in Figure 5.2 are labelled $w1$ to $w4$. The human neuron is always in one of two states. It either fires if the sum of its inputs are greater than a certain threshold, or remains inactive if the sum of its inputs is below that threshold. In this artificial network, however, the neuron becomes active to an extent which gradually increases as the sum of its inputs increases.
3. The human neuron normally has only one output channel or 'axon', but the artificial neurons can stimulate any number of other neurons.

Taking (1) and (2) into account, the degree of stimulation produced by an artificial neuron depends on the strength of both its inputs and its weight. The Federalist network was trained by a process called the conjugate gradient method. Starting with random weights, the network was fed data pertaining to the occurrence of the marker words in a text of known authorship, and the network would select either Hamilton or Madison as the most probable author. According to whether the network was correct or not, the weights were adjusted, then the network was given the marker-word data from another text. This process continued until the network had been 'shown' 100 documents. The weights were then frozen, and the testing phase began. The 12 disputed papers were presented to the network in turn, and in each case they were classified as being by Madison.

Holmes and Forsyth (1995) also studied the Federalist papers, using genetic algorithms which, like neural networks, are inspired by biology. Within the genetic algorithm, a set of rules, in this case a set of discriminants for distinguishing the work of different authors, is likened to a set of biological genes. The system starts with a random set of rules. During the training phase of the algorithm, the system attempts to determine the authorship of the non-disputed papers. The rule sets which lead to successful discrimination of texts are allowed to survive, while those producing incorrect decisions are allowed to die out. The surviving rule sets 'breed' by producing 'children', which take some genetic material (rules) from each parent to form new rule sets, and the process begins again. Spontaneous mutations of rules are also permitted. Finally, by a process allied to 'survival of the fittest', an optimal set of rules will emerge. Holmes and Forsyth's initial set of rules were based upon the frequencies of 30 function words, but the training phase of the algorithm pruned this number down to eight. In the testing phase, this reduced set of discriminant rules correctly classified all 12 disputed papers as being written by Madison.

2.7 Kenny's work on the Aristotelean *Ethics*

Three books now thought to have been written by Aristotle appear in both the *Nicomachean Ethics* and the *Eudemian Ethics*. A problem of interest is therefore to decide to which set these books most properly belong. Kenny (1977) used various techniques, many based on Ellegård's methods, to tackle this problem. Using word-frequency counts, Kenny showed that the disputed books were much closer to the *Eudemian Ethics*. Using a list of 36 common words, in 34 cases no significant difference was found between the disputed books and the *Eudemian Ethics*, while for 20 out of the 36 words, significant differences were found between their usage in the disputed texts and their usage in the *Nicomachian Ethics*. Similar results were obtained from the examination of prepositions and adverbs. A number of adverbs in particular occur much more frequently in the *Nicomachean Ethics* than in the *Eudemian Ethics* or the disputed books. In a later experiment, Kenny (1988) divided the disputed books into samples of about 1000 words each. Groups of common words were found which had a tendency to occur more frequently in either the *Nicomachean* or *Eudemian* ethics. Using the method of Ellegård, a distinctiveness ratio (frequency in the *Nicomachean* texts divided by frequency in the *Eudemian* texts) was calculated for each group of words. In every case, the 1000-word samples from the disputed text more closely resembled the *Eudemian* than the *Nicomachean* texts (Hockey 1980).

2.8 A stylometric analysis of Mormon scripture

Holmes (1992) determined Yule's K characteristic and other measures of vocabulary richness for 17 samples of Mormon scripture, three samples taken from the Old Testament book of *Isaiah* and three samples of Joseph Smith's personal writings. The K values were converted into a similarity matrix using the following formula for each pair of text samples:

$$S_{rs} = 1 - \left(\frac{X_r - X_s}{range} \right)^2$$

where S_{rs} is the similarity between text samples r and s, X_r and X_s are the K characteristics for samples r and s, and the *range* is the difference between the highest and the lowest K values found in the entire set of text samples. A dendrogram was produced from the similarity matrix using single linkage clustering. The results suggested that the *Isaiah* and Joseph Smith samples were relatively distinct from the samples of Mormon scripture, but the writings of different Mormon prophets were not clearly differentiated from each other.

2.9 Studies of Swift and his contemporaries

In an experiment performed by Milic (1966), seven samples of the work of Swift were taken, each sample being 3500 words long. A set of other authors were studied for comparison (Addison, Gibbon, Johnson and Macaulay), with two samples being taken from each. This corpus was part-of-speech tagged and Swift was found to be constant in his use of parts of speech, compared with the other authors. Both Macaulay and Gibbon made more use of nouns and prepositions than Swift.

More clear-cut results were obtained when sequential patterns of three-word classes (trigrams) were used. Initially it was found that the same trigrams occurred most commonly in all five writers to almost the same degree. The sequence (preposition determiner noun) was most common for all five writers. Thus, neither the most common trigram of parts of speech or its relative occurrence can be used as discriminator between the writers. The next factor to be examined was the total number of different part-of-speech trigrams found in each sample. These totals were found to be highly individual, with the highest values being found for the samples of Swift (which were consistent with each other) and thus could potentially serve as a differentiating criterion. This value was called the D value. Another criterion examined was the percentage of occasions on which each part of speech was assigned to the first word of a sentence. Swift was often found to use connectives at the beginning of sentences. He also used significantly fewer pronouns and conjunctions. Almost all the control authors used more initial determiners than Swift.

At the time Milic was writing, *A Letter of Advice to a Young Poet* was the only major work in the Swift canon to be in doubt. This letter was divided into three samples, and compared with the other Swift samples and the controls according to word-class frequency distributions. The letter samples were more consistent both with each other and the mean Swift values than they were with the controls. Since it is not known who else may have written the letter, we may infer that it is likely that Swift was the author.

In summary, the three tests found by Milic to be the most reliable discriminators between Swift and other authors were high scores for the use of verbals (VB, infinitives, participles and gerunds), introductory connectives (IC, co-ordinating and subordinating conjunctions, and conjunctive adverbs) and different three-word patterns (D, trigrams of parts of speech).

Köster (1971), like Milic, was interested in the work of Swift and other authors who worked to produce propaganda for the Tory party. One example of their publications was *The Story of the St Albans Ghost* (SAG). In an experiment to determine the authorship of this piece, known works by Swift, Arbuthnot and Wagstaffe were given part-of-speech tags according to a scheme slightly different from that of Milic. Köster also tried to use Milic's discriminators on both the known author and the disputed texts. The results obtained by both Milic and Köster are given in Table 5.7. The part-of-speech frequency distribution and the *D* score for SAG were consistent with Swift's writing, but SAG produced lower scores than Swift for both verbals and introductory connectives. Thus, Köster concluded, SAG was probably not written by Swift.

Other possible authors of SAG who have been suggested are Swift in collaboration with Arbuthnot or Wagstaffe. No samples of work known to be written by both Swift and Arbuthnot were available, but three works by Arbuthnot alone (*The Art of Political Lying* (APL), the *John Bull* pamphlets (JB) and *A Sermon Preached at the Mercat Cross of Edinburgh* (SMC)) were used. These samples showed great variability within Arbuthnot's writing – between 14.9 per cent (higher than Milic's control samples) and 29.7 per cent (within the Swift range) for introductory connectives and from 13.4 per cent to 16.2 per cent for finite verbs. In fact, Köster found no grammatical feature in which Arbuthnot was both consistent in himself and clearly differentiated from Swift.

The signed works by Wagstaffe used in the study were *A Letter to Dr Freind, Shewing the Danger and Uncertainty of Inocculating the Small Pox* (two sections, LF1 and LF2) and the preface to *Anthropologia Nova* (ANP). The samples of 'pseudo-Wagstaffe', being words probably by Wagstaffe though not signed by him, were items from the so-called *Miscellaneous Works of Dr Wagstaffe*, whose authorship is in doubt, namely *A Comment upon the History of Tom Thumb* (CTT), *The State and Condition of our Taxes* (SCT), and *Testimonies of the Citizens of Fickleborough* (TCF).

The results for Wagstaffe show consistent scores for verbals (VB), lower than those for Arbuthnot or Swift, and high scores for introductory connectives (IC), some similar to Swift and some higher. Wagstaffe's results for different (D) patterns or part of speech trigrams were less consistent than those for either Swift or Arbuthnot. The results for pseudo-Wagstaffe were highly irregular, with different results for each of the three samples.

The last author to be considered was the outside control. A sample of *Memoirs of Europe* (ME) by Mrs Manley (who wrote in the Tory in-house style of the period, but is not associated with *The Story of the St Albans Ghost*) was examined. The discriminant patterns of ME were closest to those of the *John Bull* sample. The results for each text sample discussed in this section for Milic's 'three discriminator profile' are shown in Table 5.7. In this table, GULL denotes a sample taken from *Gulliver's Travels*. Overall, Köster's results appear inconclusive.

The three discriminants of Milic resemble the types of quantitative features chosen by Biber (1995) to distinguish genres of text. There are thus similarities

	VB	IC	D
Controls			
Highest score	3.7	17.8	769
Lowest score	1.6	3.8	440
Mean	2.6	11.6	652
Swift			
Highest score	4.4	41.8	868
Lowest score	3.5	24.0	768
GULL1	3.5	24.0	789
GULL2	3.8	26.1	768
EX2	4.2	41.3	844
Mean	4.1	33.1	833
a) Results of Milic			
Swift			
EX2	3.87	41.4	844
Arbuthnot			
APL	3.94	14.9	795
JB	3.75	22.3	780
SMC	3.52	29.7	840
Mean	3.74	22.3	805
Wagstaffe			
LF1	2.54	38.3	827
LF2	2.56	42.5	861
ANP	2.64	44.1	740
Mean	2.58	41.6	809
Pseudo-Wagstaffe			
CCT	2.95	22.1	794
SCT	2.89	30.5	737
TCF	3.60	45.6	756
Mean	3.15	32.7	762
Mrs Manley			
ME	4.03	21.8	863
Unknown			
SAG	2.50	6.9	823
b) Results of Köster			

Table 5.7 Results for text samples by Swift and three contemporaries according to Milic's three discriminator profile

between the techniques used to distinguish text genres and those used to distinguish authorship styles. In their experiments, Baayen, van Halteren and Tweedie (1996) found that observable differences in text register tend to swamp those which arise due to differences in authorship (McEnery and Oakes 1997).

2.10 Was *And Quiet Flows the Don* written by Mikhail Sholokhov?

Kjetsaa (1976) and colleagues worked on the problem of whether the novel *And Quiet Flows the Don* was an original work of Mikhail Sholokhov or whether it had been plagarised from works by Fyodor Kryukov. In their initial study, six samples each of 500 random sentences were used, two each taken from novels known to be by Sholokhov and Kryukov respectively, and two taken from sections of *And Quiet Flows the Don*. Using a single value for average sentence length did not differentiate between the texts, but a comparison of the whole spectrum of sentence lengths expressed as a percentage of the total number of sentences showed that the values for Sholokhov and *And Quiet Flows the Don* coincided, while the values obtained for Kryukov remained distinct. When using the chi-square test to compare the relative usage of six parts of speech, it was possible to reject the hypothesis that Kryukov and *And Quiet Flows the Don* were taken from the same population, while this conclusion could not be reached for Sholokhov and *And Quiet Flows the Don*. They then used the position of a part of speech in a sentence as a potential discriminating criterion, since earlier studies in Russian had shown that the first two and last three positions in a sentence produced good results for this purpose. In general, since inflected languages such as Russian have a relatively free word order, tests based on word position are good discriminators of style. They also examined part-of-speech bigrams at the start of the sentence, and trigrams at the end of sentences. In each case, the style of Sholokhov closely matched that found in *And Quiet Flows the Don*, while the style of Kryukov was found to differ (Oakman 1980).

2.11 The authenticity of the Baligant episode in the *Chanson de Roland*

The *Chanson de Roland* is an 11th-century epic poem, of which versions survive in French, Latin and Norse. One section of the work is the Baligant episode concerning a battle between Charlemagne and Baligant. Allen's (1974) work in computational stylistics helped determine if the Baligant episode is stylistically consistent with the rest of the poem, and therefore probably part of the original poem, or whether it was a later addition by another author or authors.

Since spelling was not consistent in the version of the poem that Allen was working with, each word was replaced by its **lemma**. Each lemma was also classified according to one of 45 grammatical categories, so that in most cases homographs would be differently encoded. If more than one sense of a word fell into the same grammatical category, a digit was appended to the end of the word to indicate word sense. The use of lemmas meant that word-length information was lost, so word length could no longer be used as a discriminator of style. The test that was successfully used on the lemmatised text was one of vocabulary distribution. The text was divided into four roughly equal parts for comparison. Three sections were labelled x, y and z, and the fourth section was the Baligant episode. Since only high-frequency function words were to be used in this comparison, only those 16 words which made up 1 per cent or more of the text sample in two or more of the sections were retained. The most frequent lemma was the definite article *li* which, in its various grammatical and contractional forms, accounted for 5.5 per cent of the words in section x, 6.1 per cent of those in y, 5.8 per cent in z, and 7.2 per cent of the words in the Baligant episode. The following criterion was used to determine whether any of those values differed significantly from the others: if the greatest percentage difference between any three sections was less than the smallest difference between those sections and the remaining one, that word was considered to discriminate significantly. Using this criterion, *li* occurs significantly more often in the Baligant episode than in sections x, y and z. No significant difference was found in the usage of the second most frequent lemma, *e*, meaning *and*. Altogether, of the sixteen original lemmas used as potential discriminants, six did not vary significantly between the samples; six were used either significantly more or significantly less frequently in the Baligant episode; two distinguished section y and one distinguished section z. The probability that any one of four sections will be chosen in six out of nine cases is 0.01, which is regarded as being significant. These results show that the Baligant episode is stylistically distinct from the other sections, and thus may have been composed by another poet.

2.12 The chronology of Isocrates

Closely related to the problem of disputed authorship is the problem of chronology for a single author. Chronological studies, such as that of Michaelson and Morton (1976), aim to find which of an author's works were written at which stage in that author's career. Isocrates was chosen for this study of chronology, since he produced a number of literary works on various subjects over a period of 65 years. The historical background is sufficiently well known so that his works may be accurately dated. The object of the study was to find a general solution concerning what aspects of Greek prose style might be used to distinguish the work of a single author at various points in his career.

Michaelson and Morton found that the range of vocabulary used in the early part of Isocrates' sentences was restricted, while a greater choice of words was found in the latter part of the sentences. This data is summarised in Table 5.8.

They examined the occurrence of function words in the early part of sentences in their search for chronological discriminators. Their most significant

Position in sentence	Number of different words in position
1	768
2	405
3	1037
4	1437
5	1600

Table 5.8 The positional distribution of the vocabulary in the orations of Isocrates

results are shown in Table 5.9, where Y is the year of publication of the oration, counting back from 339 BC; S is the number of sentences in that oration; $K1$ denotes a sentence with *kai* (*and*) as its first word; $G2$ and $G3$ denote sentences with *gar* (*for*, conjunction) as their second or third word; and $M2$ denotes a sentence with *men* (*on the one hand*) as the second word. Other features examined but not included in the table were sentences with *de* (*on the other hand*) as the second word, sentences with *alla* (*but*) as the first word and the occurrence of three forms of the negative particle.

Y	S	K1	G2	G3	M2
66	45	5	10	2	2
65	57	3	12	5	9
64	154	18	16	7	12
59	98	11	21	3	8
56	128	17	25	8	13
55	155	20	21	5	16
52	46	6	11	4	5
50	106	11	27	5	13
42	360	33	95	22	39
36	138	9	26	11	24
34	126	13	34	9	11
32	129	14	32	14	19
31	158	8	31	8	19
30	173	17	34	16	32
28	224	12	46	15	26
17	182	16	56	12	19
17	289	15	72	22	34
16	654	41	150	41	87
8	321	18	86	12	25
1	475	22	99	37	71

Table 5.9 Occurrence of key function words in early positions in the sentences of Isocrates' orations

Michaelson and Morton made a regression analysis of the rate of occurrence of the key function words against the date of composition. From this analysis, four characteristics were found which vary linearly to a significant extent with the year of authorship. The four characteristics were: average occurrence of *kai* per sentence, proportion of occurrences of *kai* in which it is the first word of the sentence, proportion of occurrences of *gar* in which it is the second or third word of the sentence and proportion of occurrences of *men* in which it is the second word of the sentence. The correlation coefficient of these variables when plotted against time, together with the formula of the straight line which most closely fits the points on the graph of characteristic score against year of publication, are shown in Table 5.10.

Characteristic	Correlation coefficient	Estimated year of publication
Kai **per sentence**	+ 0.70	$0.82x + 88.8$
Kai **as first word**	+ 0.78	$1.33x + 27.4$
Gar **as 2nd or 3rd word**	− 0.47	$0.001x + 2.8$
Men **as 2nd word**	− 0.53	$0.06x + 12.2$

Table 5.10 Correlation coefficients and regression equations for four characteristics which show chronological change

In each case there are 18 degrees of freedom and for this the correlation coefficient is 0.44 for $p = 0.05$ and 0.59 for $p = 0.01$. The degree of variation for *gar* and *men* is too slight to be of practical value, but Michaelson and Morton estimate that the regression equations for *kai* may be used to give time estimates accurate to plus or minus four years.

2.13 Studies of Shakespeare and his contemporaries

The main difficulty with using stylistic criteria to distinguish between authors in dramatic texts is due to the following contradiction: on the one hand, a good author is likely to have a very distinctive style, while on the other hand, a good author will also exhibit a variety of styles in the range of spoken parts.

Baillie (1974) was interested in examining the possibility that *Henry VIII*, normally attributed to Shakespeare, was largely written by Fletcher, a view held by many scholars. In a pilot study, two texts known to be written by Shakespeare (*Cymbeline* and *A Winter's Tale*) and two plays known to have been written by Fletcher (*The Woman's Prize* and *Valentinian*) were used as comparison texts with ten 500-word samples being taken from each text. Using the EYEBALL program, no fewer than 65 potential stylistic discriminators were examined. Although no single discriminator was sufficient to distinguish between the two authors, it was found that when certain pairs of variables were examined simultaneously the power of discrimination was greatly increased.

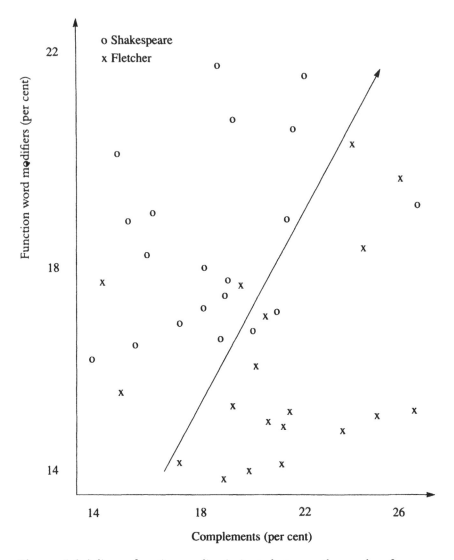

Figure 5.3 A linear function to discriminate between the works of Shakespeare and Fletcher

For example, it was possible to plot all 40 samples on a graph where the horizontal axis was the percentage of complements and the vertical axis was the percentage of function-word modifiers. This graph is reproduced in Figure 5.3.

A diagonal boundary line separated two regions of the graph: the high function-word/low complements region which consisted mainly of samples by Shakespeare, and the low function-word/high complements region which contained most of the samples by Fletcher. In this, 34 out of the 40 samples were correctly discriminated, a proportion that could be obtained by chance

Figure 5.4 Variation of Yardi's discriminant function with the date of production of Shakespeare's plays

only one time in a thousand. A similar graph where the axes were the percentage of noun modifiers and the ratio of co-ordinators to subordinators correctly classified 32 out of the 40 samples.

Williams (1970) describes how, in 1946, Yardi produced his discriminant function to determine the chronology of Shakespeare's plays. This combined into a single value the number of 'full split lines', the number of lines with 'redundant final syllables' and the number of 'unsplit lines with pauses', all in relation to the total number of speech lines. The final value of his function is the weighted sum of all these factors. These values represent the tendency to break away from formal rhythm as Shakespeare grew older. Yardi divided 30 of the plays, for which the date of production was moderately certain, into 21 successive groups of approximately one year. From these he calculated a regression line for the change in his function with time. He found an approximate straight-line relation given by $u = 0.7204 + 0.07645v$ where 0.7204 is the approximate function for the earliest plays, to which must be added 0.07645 for each year later in Shakespeare's development. The variation of Yardi's discriminant function with the date of production of Shakespeare's plays is shown in Figure 5.4.

If u, the discriminant function of a particular play, is known, an estimate of v can be found from the formula, indicating the approximate date of production in years after the earliest of Shakespeare's plays. Yardi used this method to suggest approximate dates of writing or production for several plays for which there was little external evidence for these. Thisted and Efron (1987) describe how a newly discovered poem was attributed to Shakespeare.

2.14 Comparison of Kierkegaard's pseudonyms

In previous sections we discussed the problem of a single author writing in different styles over a period of time. In this section we will examine the related problem of the same author writing in different circumstances. For example, Kierkegaard used different pseudonyms for works which, viewed individually, present alternative points of view. The work produced under each pseudonym is internally consistent with its own distinctive point of view. McKinnon and Webster (1971) describe a method which has served to establish a hierarchy among eight of Kierkegaard's most important pseudonyms. Kierkegaard himself pointed out that that there were contradictions between the views expressed by different pseudonyms – contradictory quotations can be found among them. McKinnon and Webster's task was to produce a method to determine which of Kierkegaard's pseudonyms are closest to his attributed work.

Altogether, 16 sections of text were used for this study, with eight sections of acknowledged work and eight sections of pseudonymous work. Three vocabulary lists were produced for each sample of text: total vocabulary, vocabulary exclusive to that section and the vocabulary in each of the pseudonymous sections shared exclusively with the acknowledged sections. The experimental procedure consisted of two stages:

    1a. a comparison of the text densities (related to the type-token ratio) of the pseudonymous (PS) and acknowledged (SK) samples

    1b. a comparison of their internal coherence or homogeneity by the vocabulary connectivity method

    2.   a pair vocabulary test, the purpose of which was to establish a hierarchy of the synonyms in relation to the acknowledged Kierkegaard.

The purpose of the first two tests was to show whether PS differs significantly from SK.

V was used to denote the total vocabulary size, and N the total number of words in the sample (including repetitions). Test 1a used the measure $logV/logN$, since it had been found empirically that vocabulary size grew with text length according to a bilogarithmic relationship. This ratio was 0.819 for SK and 0.845 for PS, showing PS has the richer vocabulary. This supported the hypothesis that PS is significantly different to SK.

The same result was obtained for the vocabulary connectivity method. This involved a comparison between observed vocabulary exclusive to a given section (O) and the calculated values for a chance distribution taking into account text length and rank frequency distribution of each selection (C). McKinnon and Webster do not give details of the formula used for these calculations. O tended to be greater than C for PS, and took a greater variety of values, while O was approximately equal to C for SK. This shows that SK was much more homogeneous than PS, and PS had a richer vocabulary. The various pseudonyms were ranked by the O/C ratio. The closest to the acknowledged works was *Sickness Unto Death*, by 'Anti-Climacus', while the furthest was *Repetition* by 'Constantine Constantius'.

2.15 Dogmatic and tentative alternatives
Another example of variation within the work of a single author is in the use of

dogmatic and tentative alternatives, as found by Ertel. Examples of dogmatic words are *always, whenever, never*, and *absolutely*, while examples of tentative expressions are *often, sometimes, occasionally* and *up to a point*. It was possible to define a dogmatism quotient which varies between authors, being, for example, high in Marx but low in Russell. A dogmatism quotient can also vary within the same author at different periods in time. For example, Hitler's speeches were more dogmatic in times of crisis, and less so in times of victory (Kenny 1982).

2.16 Leighton's study of 17th-century German sonnets

Leighton (1971), in a study of 17th-century German sonnets, attempted to produce a type of literary geography, where different towns might be distinguished by the style of sonnets produced there. A manual coding system was used to indicate the stylistic features of interest in each line, where the following codes were used: main clause, interrupted main clause, completion of main clause, elliptical main clause and extension phrase in apposition. Question marks, apostrophes, end of line and end of sentence were also encoded. Once this coding had been completed, various computer analyses were possible, such as the average number of main clauses per poem, average number of sentences per poem, and the commonest code patterns for each poet.

Leighton used an analysis which distinguished between main and subordinate clauses. Code *b* was used for an interrupted main clause, while *i* was used for a subordinate clause. For example, line 1 of the poem *Lambs that learn to walk in snow* by Philip Larkin would be encoded *bi*, with the phrase *that learn to walk in snow* being counted as a subordinate clause. The material chosen for the initial experiments consisted of 20 religious sonnets by Fleming and Gryphius's 31 so-called *Lissa* sonnets. It was shown that each poet had preferred line patterns, as shown in Table 5.11.

	Fleming	Gryphius
Main clauses per sentence	1.65	2.62
Sub clauses per sentence	0.88	3.57
Main clauses per poem	13.45	7.25
Sub clauses per poem	7.20	9.90
Sentences per poem	8.15	2.77
Coincidences	6.15	2.70
Enjambements	1.95	2.80
Apostrophes	2.40	1.10
Extension phrases	2.40	1.93
Percentage of octet divisions	80.00	32.26

Table 5.11 Mean frequencies of standard features in sonnets by Fleming and Gryphius

The second stage of the experiment was to produce a program to provide for any given group of sonnets a ranked frequency list of line patterns. The most important line pattern frequencies were *as* (main clause, end of line and sentence) for Fleming and *ir* (subordinate clause, end of line) for Gryphius. The range of different line patterns used was greater for Fleming, who used 151 line patterns in 20 sonnets as compared with Gryphius who used only 128 in 31 sonnets.

2.17 Forensic stylometry

Morton (1978) describes a technique for discriminating between real and fabricated confessions where authentic material is provided to represent the accused. This can take one of two forms. Ideally, another statement made by the same person will be used, but if evidence of this type cannot be produced, the accused is asked to write an essay on any neutral subject such as 'My Schooldays'. It is also vital to know the circumstances in which the author produced a text, particularly when determining the authenticity of confessions made in police custody and used as evidence in trials. Morton asks, if a confession were made under pressure, would the prisoner have used his own natural style or would he assume the language of his captors? Morton achieved fame as an expert witness, showing in court that of a set of 11 statements allegedly made to the police by one author, four were written by a different author. The accused was subsequently acquitted of charges based on these four statements.

Morton's work has been accepted in British courts, but not in the USA because of such problems as differences between letters, diaries and oral confessions by the same person being likely to be greater than differences between two different people writing in the same genre. For example, a comparison between statements made before the Hearst trial and tape-recorded statements made by members of the Symbionese Liberation Army was not accepted in court (Kenny 1982). Other examples of forensic stylometry include determining the authorship of contracts and wills, the authorship of anonymous letters in serious crimes and the authenticity of suicide notes. If forensic stylometry develops to a sufficient extent, we may encounter the notion of stylistic 'fingerprints' in the future. However, to date, more success has been found in determining the constancy of style between texts written by the same author than the uniqueness of an individual's writing style (Kenny 1982).

One such controversial method is the **cusum** technique. This is based on the premise that everyone has a unique set of quantifiable linguistic habits, such as the frequent use of very short words or words beginning with a vowel, and the occurrence of any particular habit is examined with each cusum test. First, a cusum plot is produced, derived from the sentence lengths in the text under scrutiny. The mean sentence length is found, and then the difference between the actual and expected numbers of words found for each sentence is plotted. For example, if the mean number of words per sentence is 10, then the expected number of words in the first three sentences would be 30. If the actual number of words in the first three sentences were 5, 7 and 9, the total would be 21. The difference between the expected and actual number of words in the first three sentences would be 30 − 21 = 9, and this is the value that would be plotted on the cusum chart for sentence number three. The same procedure is used to plot the difference between the expected and actual occurrences of the linguistic habit forming the test criterion, up to and including each sentence point. If the cusum plots of sentence lengths and habit words follow each other closely, the test suggests that the text was written by a single author, while if the two plots diverge greatly, the test suggests that more than one author was responsible. If cusum charts reveal that a confession has been written by more than one author, this of course suggests that material not in the linguistic style of the author of the original confession has been inserted later. Canter (1992) found that the test was not reliable, whether one relied upon a subjective appraisal of a visual comparison of the two cusum plots, or attempted to quantify the correlation between the two lines using Spearman's rank correlation coefficient. A review of the controversy over the cusum technique, which looks at a number of modifications of the basic technique, has been written by Holmes and Tweedie (1995).

2.18 The use of a syntactically annotated corpus in stylometry

Baayen, van Halteren and Tweedie (1996) performed experiments in authorship attribution which made use of a syntactically annotated corpus. Statistical measures previously applied to words and their frequencies of use were applied in a similar manner to syntactic phrase rewrite rules as they appeared in the corpus. In various authorship studies, the words which seem to be the best discriminators between the work of different authors are the so-called function words such as *a*, *the* and *that*. The use of these words reflects the underlying syntax of the text in which they occur. This suggests that a direct study of syntax use might yield results at least as good as those obtained by the statistical analysis of word frequencies in terms of their discriminatory potential. Furthermore, there is some evidence (such as that found by Baayen, van Halteren and Tweedie in the pilot study described here) that the frequencies of rewrite rules are less subject to variations within the text than are word frequencies.

In a pilot study, the authors found the 50 most frequent words in the Nijmegen corpus[1] as a whole. These texts were given sets of scores using the technique of principal components analysis described in Chapter 3, Section 2.1. The first three components extracted accounted for 52 per cent of the variance, and the scores obtained by each text on these three components were displayed on three scatter plots, as described by Horvath in Chapter 3, Section 2.5. The scatter plot for principal components 1 and 2 visually grouped together texts from similar genres: one cluster corresponded to drama, one to crime fiction and one to scientific texts. However, the scatter plots for principal component 3 (which provided just 9 per cent of the variance between texts) revealed author-specific differences between texts of a single genre. This suggested that genre differences are somewhat greater than author differences.

Baayen, van Halteren and Tweedie used two texts of the same register (crime fiction) taken from the Nijmegen corpus. These texts were chosen so that no effects of difference in register would cloud the effects of different authorship, and the texts were also syntactically annotated with the TOSCA system. For the actual experiments, they used only the crime fiction novels of Allingham and Innes, taking a sample of 20,000 words from each author. Both samples had been syntactically annotated with the TOSCA annotation scheme, which consists of

- the syntactic category, i.e. the general nature of the constituent itself,
- the syntactic function, the role the constituent plays within a larger constituent, and
- additional attributes of interest such as 'singular'.

For example, the sentence *He walks his dog in the park* is annotated by a single tree structure, where the leaves correspond to single words, which are grouped under other labels corresponding to successively longer phrases. Within that tree, the noun *park* is given the syntactic category *N*. Its syntactic function, namely the head of the noun phrase *the park*, is denoted NPHD. The other attributes of interest are labelled *com* (common) and *sing* (singular). The labels for the individual words *the* and *park* are combined under the label PC, NP, which shows that *the park* is a noun phrase which functions as a prepositional complement. This tree structure can be represented by rewrite rules, where, for example, *the park* would be represented by the rewrite rule PC:NP > DT:DTP + NPHD:N, where DT:DTP denotes that *the* is a determiner phrase which functions as a determiner. After counting the total number of rewrite rules needed to describe each text and the degree of repetition of each rule, it was found that the resulting type-token ratio was similar to that expected for the words, about 4000. The frequency of each rewrite rule was then found for both samples of text.

Baayen, van Halteren and Tweedie examined five different measures of vocabulary richness, normally used for evaluating texts on the basis of the word frequencies they contain. One of these measures was Yule's characteristic K, described in Section 2.3. The other measures were as follows:

Simpson's D, given by the formula

$$D = \sum_{i=1}^{v} V(i,N) \frac{i(i-1)}{N(N-1)}$$

where N is the number of tokens, $V(i,N)$ the number of types which occur i times in a sample of N tokens, and v the highest frequency of occurrence.

Honoré's measure gives more weight to the low-frequency end of the distribution, including the hapax legomena, denoted $V(1,N)$

$$R = 100 \frac{\log_e N}{1 - \frac{V(1,N)}{V(N)}}$$

Sichel's measure also takes into account the low-frequency words:

$$S = V(2,N)/V(N)$$

Brunet's measure of vocabulary richness is given by the formula

$$W = N^{V(N)-a}$$

where a is a constant. If a is set to 0.17, W will be relatively independent of N.

The crime fiction texts were divided into 20 samples, 14 labelled with the author's name (7 each by two different authors), and six test samples, where the identity of the author was kept secret from the experimenters. Using the frequencies of the 50 most frequent words in the pooled set of all the samples, values for the five measures of vocabulary richness were obtained for each of the 20 text samples in the experiment. Thus, 20 observations were made for each of the five measures. These results were input to a principal components analysis, which combined W and R in the first principal component, and combined K and D in the second. Each of the 20 texts was given scores on both of these principal components, enabling them to be viewed on a scatter plot. This showed that when word frequencies were used as the basis of the discrimination test, only four out of six of the test samples were classified correctly.

A similar experiment was then performed to find out whether the 50 most frequent rewrite rules provided a better basis for author discrimination than the 50 most frequent words. This time the vocabulary richness measures were used on the frequencies of the 50 most frequent rewrite rules, and the resulting data input into a principal components analysis. This resulted in all six test samples being correctly clustered with samples known to be by the same author. Similarly encouraging results were obtained when the discriminatory potential of the lowest-frequency rewrite rules (especially the hapax legomena) were examined. Baayen, van Halteren and Tweedie concluded that the frequencies with which syntactic rewrite rules are put to use provide a better clue to authorship than word usage.

The fact that differences in genre were found to be easier to identify than differences in authorship helps to explain the success of Biber (1995) in his work on differentiating texts on the basis of genre or register.

3 STUDIES OF LINGUISTIC RELATIONSHIP
3.1 Ellegård's method for the statistical measurement of linguistic relationship

Ellegård (1959) invites us to assume that there once was a language L consisting of just 20 roots. This language gave rise to four daughter languages, called A, B, C and D. Languages A and B are closely related, while C and D are remote from both A and B and also from each other. We can say this because A and B have many roots in common, while C and D have few roots in common. All four languages have lost some of the roots originally present in L. It must be assumed that the vocabulary list for language L is no longer known. This means that in order to estimate which roots were once in L, one can assume that any root surviving in more than one modern language must have been present in L. If this assumption is made, we can use roots 1–13 of the original 20, as shown in Table 5.12.

Root	L	A	B	C	D
1	+	+	+	−	−
2	+	+	+	−	−
3	+	+	−	−	+
4	+	−	+	+	+
5	+	+	+	−	−
6	+	+	−	−	+
7	+	+	+	−	−
8	+	−	+	+	−
9	+	+	+	−	−
10	+	+	+	−	+
11	+	−	+	+	−
12	+	−	−	+	+
13	+	−	−	+	+
14	+	−	−	−	+
15	+	−	−	−	+
16	+	−	−	+	−
17	+	−	+	−	−
18	+	−	−	−	+
19	+	−	−	−	−
20	+	+	−	−	−

Table 5.12 Occurrence (+) and loss (−) of 20 L-roots in the daughter languages A, B, C and D

To estimate the closeness between any two of the modern languages such as A and B, we first construct a contingency table where a is the number of times the original root occurs in both A and B (6), b is the number of times the original root is found in A but not in B (2), c is the number of times the original root is found in B but not in A (3) and d is the number of times the original root is found in neither of the daughter languages (2). The product–moment correlation coefficient (r) can then be measured according to the following formula:

$$r = \frac{ad - bc}{\sqrt{(a+b)(a+c)(c+d)(b+d)}}$$

which, in the case of A and B,

$$r = \frac{12 - 6}{8 \times 9 \times 5 \times 4} = 0.16$$

r varies in the range -1 (for a pair of languages which are not related at all) to $+1$ (for a pair of languages which have all roots in common).

If the roots of the original language L are still known, all 20 rows of Table 5.12 may be used to calculate the correlation coefficient between each pair of languages. A comparison of r', the correlation for the sample of 13 roots with r'', the 'true' correlation for the whole population of 20 L-roots, is given in columns (a) and (b) of Table 5.13.

Pair	(a) r'	(b) r''	(c) r_n'	(d) r_n''
AB	0.16	0.30	0.71	0.62
AC	−1.00	−0.59	0.00	0.00
AD	−0.22	−0.21	0.43	0.33
BC	−0.16	0.00	0.45	0.39
BD	−0.72	−0.50	0.27	0.21
CD	0.22	0.07	0.55	0.40

Table 5.13 Correlation values for a set of four daughter languages of L, calculated according to whether or not the roots of L are known, and using two different formulae

Columns (a) and (b) show that whereas r' places the CD correlation higher than the AB correlation, r'' places the AB correlation higher than the CD correlation. The problem with the r', measure is that the correlation coefficient r gives equal weight to both negative agreements ($- -$) and positive agreements ($+ +$). However, a positive agreement between two languages is in general a much rarer and more significant event than a negative disagreement. Also, when the set of roots in L is not known, there is a danger of either overestimating or underestimating the number of negative agreements. A and B agree for most roots, and all their positive agreements will be included in the sample, while most of the negative agreements will be left out except in those relatively rare cases where they coincide with a positive agreement for C and D. For C and D, many negative agreements will be included, whenever they correspond with positive agreements for C and D. Thus, when estimating correlation among daughter languages when the roots of the original language are not known, negative agreements should be given less weight than the positive agreements and disagreements. This can be done by defining a new correlation coefficient r_n, where only positive agreements appear in the top line of the equation, as shown below:

$$r_n = \frac{a}{\sqrt{(a+b)(a+c)}}$$

A value of r_n near zero shows a lack of relationship, while a value near $+1$ implies a close relationship. Values of $r_{n'}$ (r_n for the sample of 13 roots where the roots of L are unknown) and $r_{n''}$ (r_n for the full population when the 20 roots are known) are shown in columns (c) and (d) respectively of Table 5.13. Both measures give the same rank ordering of closeness of the language pairs.

More closely related pairs of languages have probably separated from each other and from the original language L more recently than less close language pairs, which separated from each other and from L much longer ago. r_n was applied to a comparison of Indo-European language subfamilies. These values of r_n, expressed as percentages, are given in Table 5.14.

	Celto-Italic	Greek	Armenian	Indo-Iranian	Slavo-Baltic	Germanic
Greek	67					
Armenian	46	49				
Indo-Iranian	63	64	45			
Slavo–Baltic	65	63	43	59		
Germanic	71	63	44	62	71	
Albanian	40	44	36	41	37	38

Table 5.14 Modified correlation coefficients, expressed as percentages, for seven Indo-European subfamilies

Thus, Albanian and Armenian are most distinct from each other and from all the other language groups, while Germanic, Slavo–Baltic and Celto-Italic have the greatest number of roots in common.

3.2 Language divergence and estimated word retention rate

Dyen, James and Cole (1967) obtained estimates of language divergence and word retention from determinations of the number of cognate words found in pairs of Austronesian languages. For example, the meaning *five* is expressed as *lima* in Malay and *lima* in Tagalog, so we have found a pair of cognate words. *Six* is *enam* in Malay and *anim* in Tagalog, so this pair is also probably cognate, being derived from a common root. However, *seven* is *tujuh* in Malay and *pitu* in

Tagalog, so this word pair is not cognate. Altogether, they considered 196 different meanings in the 89 languages listed by Dyen in his 1962 study. These 196 meanings were taken from a list of 200 prepared by Swadesh (1952) from which the meanings *freeze, ice, snow* and *that* were excluded.

Different retention rates are found in words of different meanings. This means that the words for such concepts as *five* or *mother* and *father* remain constant in a given language for much longer than those describing other concepts, and hence are more likely to be cognate with the words for the same concept in a related language. Each meaning in Swadesh's test list is assumed to have a time constant τ which measures the retention rate of the words which are listed for that meaning. Each pair of languages is assumed to have a certain time separation *t* which is taken to be twice the time that has elapsed since they belonged to a common antecedent and were identical. The reciprocal 1/*t* is a direct measure of relationship between languages. We will see later how *t* and τ correspond to physical time in years.

Dyen, James and Cole state that cognation in two languages is a random phenomenon with the probability assigned by the exponential holding model which is used to describe radioactive decay. Just as the decay of a single radioactive atom is a random event, so is the change in a word of a language which means that the word is no longer cognate with its counterpart in another language. Using the exponential holding model, the probability that the *i*th pair of languages has a pair of cognate words for the *j*th meaning is given by the expression

$$p_{ij} = \exp(-t_i/\tau_j)$$

The proportion of cognate pairs found for a given meaning in all language pairs is called the productivity *P* of that meaning. The productivities of certain words in the Swadesh list in Austronesian languages is given in Table 5.15. The time constant τ is a monotonically increasing function of the percentage cognation. In other words, high productivity corresponds to a high constant τ, low productivity to a low value of τ.

Rank	Meaning	*P*
1	five	80.5
2	two	78.9
3	eye	77.2
4	we	74.7
5	louse	71.2
10	four	55.7
20	name	34.4
50	right (hand)	14.4
100	to count	5.9
196	to play	0.7

Table 5.15 Productivity percentages of lexicostatistical list meanings

In order to estimate the relationship between the productivity and τ, meanings were grouped into nine classes of approximately equal size on the basis of their productivity over all pairs of the 89 lists. The 22 most productive meanings were grouped into class 1, the next 22 into class 2 and so on. For each pair in a subsample of 46 language pairs, the proportion of cognates was found in each of the nine classes of meanings. As an example, the results for one language pair (Tagalog and Ratagnon) are presented in Table 5.16. From the relation

$$p_{ij} = \exp(-t_i/\tau_j) \text{ we obtain}$$

$$\log_e p_{ij} = -t_i/\tau_j \text{ and}$$

$$\log_e(-\log_e p_{ij}) = \log_e t_i - \log_e \tau_j$$

$\log_e(-\log_e p_{ij})$ is called the **log log transform** of p_{ij}. The additivity of the effects t_i and τ_j on the log log transforms of the true probabilities p_{ij} suggests that we could obtain estimates of t_i and τ_j by taking the log log transforms of the observed proportions $\hat{p}_{ij}$ which approximate p_{ij}. The notation $\hat{p}_{ij}$ indicates an estimation of p_{ij}. Another example of this 'circumflex' notation is found in signal processing, where s is the transmitted signal and $\hat{s}$ is the received signal. At various points across the signal route, noise corrupts the waveform s. Since we do not know the extent to which the received signal has become corrupted by noise, it gives us only an estimate of the transmitted signal (Sklar 1988). The log log transforms of $\hat{p}_{ij}$ given by $\log_e(-\log_e \hat{p}_{ij})$, are given in Table 5.16.

Meaning class	1	2	3	4	5	6	7	8	9
% Cognation	90.5	86.3	65.0	63.2	50.0	47.6	52.6	27.3	22.0
log log transform	−2.30	−1.92	−0.84	−0.78	−0.37	−0.30	−0.44	−0.26	−0.41

Table 5.16 Percentages of cognation $\hat{p}_{ij}$ and their log log transforms in each productivity class for the list pair Tagalog and Ratagnon

1. The total of log log transforms over all meaning classes for Tagalog and Ratagnon is −6.28, with a mean of −0.70.
2. Although there is insufficient space to tabulate the full data set here, the total of log log transforms over all language pairs for meaning class 1 was found to be 77.2 with a mean of −1.68.

In the following discussion, the log log transform of $\hat{p}_{ij}$ will be denoted x_{ij}. The average of x_{ij} over the nine classes over which the subscript j runs is denoted by $\bar{x}_i$. The dot in $\bar{x}_i$ indicates the subscript with respect to which we have averaged. The means $\bar{x}_j$ and $\bar{x}_i$, given in Tables 7.17(a) and 7.17(b) respectively, are calculated in the same way as the quantities labelled 1. and 2. just below Table 5.16. $\bar{x}_j$ is the total of the log log transforms over all meaning classes for a given language pair divided by the number of classes (9), while $\bar{x}_{i.}$ is the total of the log log transforms over all language pairs for a given meaning class, divided by the number of language pairs (46).

If we average x_{ij} with respect to j and i in turn, we obtain

$$x_{i.} = \log_e \hat{t}_i - \overline{\log_e \tau} \quad \text{(equation i)}$$

$$x_{.j} = \overline{\log_e t} - \log_e \hat{\tau}_i \quad \text{(equation ii)}$$

We can standardise our time measures by defining the unit of t and t to be the time constant of the ninth meaning class. Thus, if we set $\hat{\tau}_9 = 1$ then $\log_e \hat{\tau}_9 = 0$. Substituting $j = 9$ in equation (i), we have:

$$x_{.9} = \overline{\log_e t} - \log_e \hat{\tau}_9 = \overline{\log_e t}$$

and substituting in equation (ii) we have

$$\log_e \hat{\tau}_j = \bar{x}_{.9} - \bar{x}_{.j}$$

$\bar{x}_{.j}$ is given in Column 2 of Table 5.17(a), so we can calculate $\log_e \hat{\tau}_j$ which goes into Column 3 of Table 5.17(a), and $\hat{\tau}_j$ is column 4. From Table 5.17(a) we have $\log_e \tau = .797$. Hence from equation (ii) we have

$$\log_e \hat{t}_j = \bar{x}_{i.} + \overline{\log_e \tau} = \bar{x}_{i.} + .797$$

The values of $\log_e \hat{t}_i$ in the third column of Table 5.17(b) are thus obtained from the means $\bar{x}_{i.}$ in the second column.

The data permits the estimation only of relative times and therefore does not make possible the estimation of t_i or t_j in years or millennia. In order to estimate the absolute magnitude of the time unit, we need historical data. Dyen, James and Cole use the figure given by Lees (1953) of 81 per cent retention for a language during a millennium. This means that the average cognation for two languages (p) will be the square, that is, $p = (0.81)^2$.

Meaning class	Means ($\bar{x}_j$)	$\log_e \hat{\tau}_j$	$\hat{\tau}_j$
1	−1.68	2.04	7.67
2	−1.18	1.53	4.64
3	−0.74	1.10	3.00
4	−0.54	0.90	2.45
5	−0.17	0.53	1.70
6	−0.16	0.52	1.69
7	−0.05	0.41	1.51
8	0.21	0.14	1.15
9	0.36	0.00	1.00
Grand mean	−0.44	0.797	—

Table 5.17(a) Time constants $\hat{t}_i$ for each meaning class

Language pair	Means ($\bar{x}_i$)	$\log_e \hat{t}_i$	$\hat{t}_i$
Dibabaon x Cuyunon	0.11	0.91	2.48
Tongan x Atiu	−0.22	0.58	1.79
Tikopia x Kapingamarangi	−0.59	0.20	1.22
Samoan x Ellice	−0.79	0.01	1.01
Kantilan x Cebuan-Visayan	−1.12	−0.33	0.72

Table 5.17(b) Time separations for selected language pairs

Using the formula $\log_e \hat{\tau}_9 = \overline{\log_e t} - \log_e (-\log_e(.81)^2)$, they estimate the time unit of t and τ which was defined as the average time constant of the ninth class, to be 1069 years, so our time units are roughly millennia. The practical use of cognates in multilingual corpus-based computational linguistics has been described by Simard, Foster and Isabelle (1992) and McEnery and Oakes (1996). An account of their work is given in Chapter 3, Section 4.9.3.

4 DECIPHERMENT AND TRANSLATION
4.1 The decipherment of Linear B
Between 1899 and 1935 the archaeologist Arthur Evans excavated the city of Knossos in Crete. Among his findings were a hoard of inscribed tablets, dried but not baked. There were two types of writing on the tablets, **Linear A** which had been found at other locations in Crete, and **Linear B** which was found only at Knossos. Linear B appeared to be more recent than Linear A and, originally, was not considered to be Greek, since the ethnic origin of the Cretans was widely believed to be non-Greek (Kahn 1966). Linear B was successfully deciphered using only hand and eye. Considering the efficacy of that technique in deciphering an unknown language, it seems at least possible to allow a computer to use similar techniques in order to translate languages. We will look at work related to this in statistical machine translation in Section 4.3. However, the use of observation and human intuition is still an important part of corpus linguistics.

There are some resemblances between an unreadable natural language script and a secret code, and similar methods can be employed to break both (Chadwick 1958). The differences are that (a) the code is deliberately designed to baffle the investigator, while the script is only puzzling by accident and (b) the language underlying the coded text is normally known, while in the case of a script there are three separate possibilities. Firstly, the language may be fully or partially known but written in an unknown script. Secondly, as is the case for Etruscan, the script may be known but the language unknown. Thirdly, we have the situation which originally faced any would-be decipherer of the Linear B script, where both the script and the language were unknown. Although it was later discovered that the underlying language was known, that fact could not be used when work on decipherment first commenced. The success of any decipherment depends upon the existence and availability of adequate material. In cases where both the language and the script are unknown, a bilingual text is generally required before decipherment is possible. If as in the case of Linear B, no bilingual text is available, a far larger corpus of text is required.

The first step in deciphering an unknown script is to determine whether the writing system used is ideographic, syllabic or alphabetic. All known writing systems use one or a combination of these three basic methods. Ideographic writing provides a picture or ideogram to denote an entire concept. It requires a huge number of signs to cover even a simple vocabulary, and generally gives no guide to pronunciation. Chinese script is an example of largely ideographic writing, although some characters are made up of combinations of two or more simpler characters, one of which may provide a clue to pronunciation. A single digit such as 5 is an example of an ideogram, since it represents a whole concept without containing any of the constituent characters of *five*. The syllabic and alphabetic systems are both made up of elements which combine to represent the sound of a word. A syllabic alphabet consists of about 50 characters if the language consists of open syllables (combinations of one consonant and one vowel), but more if it contains complex syllables (containing consonant or vowel clusters as in *strength*) as is the case for English. Alphabetic systems generally do not have more than the 32 characters found in Russian. Thus, the number of characters in an alphabet gives a strong clue as to whether a writing system is ideographic, syllabic or alphabetic. Since 89 characters were originally found in samples of Linear B, it is most probably a syllabic script. Another clue is that vertical bars appear between groups of two to eight characters in Linear B, which would be expected if these groups corresponded to the number of syllables in a word. These elementary observations were disregarded during many early attempts at decipherment. However, the Linear B script also has a number of commonly occurring ideograms, consisting of pictures of tripods, and jars or cups, together with metric signs and numerals.

Alice Kober (1945) tried to discover whether Linear B was an inflected language which used different word endings to express different grammatical forms of a word. In particular, she investigated whether there might be a consistent means of denoting a plural form or distinguishing genders. At various points on the Linear B tablets, it is clear that summation is taking place, yielding totals of certain objects. Kober showed that the totalling formula had two forms according to the ideographs contained within: one was used for men and for one class of animals; the other was used for women, another class of animals, and also for swords, giving strong evidence for the distinction of gender. Kober also demonstrated that certain words had two variant forms, which were longer than the basic form by one sign. She interpreted them as further evidence of inflection; but they were destined to play an even more important role in the final decipherment (Chadwick 1958).

After recording all the common suffixes, and assigning each a code number, Kober found several nouns that were declined in three cases (Kahn 1966). Two such patterns took the format given in Table 5.18, where J K and L M indicate the respective word stems:

Case I	J K 2 7	L M 36 7
Case II	J K 2 40	L M 36 40
Case III	J K 59	L M 20

Table 5.18 Declensions in Linear B

Since there were some resemblances between Linear B and the known Cypriot syllabary, Kober conjectured that the Linear B symbols could represent either lone vowels or a combination of consonant followed by vowel, but no other combinations. Assuming that the word stems ended in consonants, which they tend to in most languages, then 2, 59, 36 and 7 might be 'bridge' signs, consisting of the last consonant of the stem and the first vowel of the ending. If so, then the pair 2 and 59 would start with the same consonant, as would the pair 36 and 20. Kober illustrated this principle with an Akkadian noun *sadanu*, whose stem is *sad-* and whose case endings are *-anu*, *-ani* and *-u*, as shown in Table 5.19.

Case I	J K 2 7
	sa da nu
Case II	J K 2 40
	sa da ni
Case III	J K 59
	sa du

Table 5.19 Declensions of the Akkadian noun *sadanu*

Although Kober did not suggest that the example given in Table 5.19 gives the actual values of the Linear B symbols, she had shown that some signs shared a common consonant. Using similar reasoning, it could be ascertained that symbols 2 and 37 share the same vowel. Since their stems are different, these two symbols probably do not commence with the same consonant, but if their case endings correspond, they will terminate with the same vowel. Kober thus discovered that some symbols had consonants in common and some had vowels in common. This enabled the identification of groups of four related symbols; they could be arranged in a square, with symbols containing the same vowel being placed in the same row and symbols sharing the same consonant being placed in the same column, as shown in Table 5.20 below:

	Consonant 1	Consonant 2
Vowel 1	2	36
Vowel 2	59	20

Table 5.20 A group of four related symbols

Bennett provided reliable lists of the signs for the first time; previous attempts had confused groups of similar signs. This enabled Ventris to compile statistical data such as the overall frequency of each sign, and its frequency in initial, final and other positions in the sign groups (see Chadwick 1958). If words are written in a syllabic script which has signs for pure vowels (vowels occurring alone and not in conjunction with a consonant), then a pure vowel symbol will only be used in the middle of a word if it immediately follows another vowel. The frequencies of pure vowel signs will then show a characteristic pattern, where the pure vowels occur rarely in the middle of a word but frequently at the beginning, because every word beginning with a vowel must begin with a pure vowel sign. This reasoning enabled identification of the symbols denoted by 08 and 38 as plain vowels. The pattern for the other pure vowels was less clear because of the occurrence of diphthongs. Ventris deduced that the symbol 78, which commonly occurred in the final position, was probably a conjunction meaning *and*, and a suffix to the word it served to connect. The fact that it was not part of the root word but a separable suffix emerged from the comparison of similar words, where one variant would be found with symbol 78 at the end, and one would be found without. Similarly, separable prefixes could be identified by observing word variants where one began with the prefix symbol and one consisted of the unprefixed root word. Further deductions were made possible by certain words which appeared in two different spellings. If they were long enough, but differed in only one syllable, and occurred several times, then there was a reasonable assumption that they had something in common. They might be variant spellings of the same word, meaning that the symbols which differed would represent similar sounds. A table of these was eventually produced. Adopting the phonetic pattern of Kober, where the symbols all represented lone vowels or combinations of consonant plus vowel, Ventris produced a grid with vowels along the top and consonants down the side. The entry in the grid corresponding to consonant III and vowel II would be the symbol which stood for the combination of these two sounds. Ventris's task was to fill in this grid with the appropriate symbols and identify the consonants and vowels corresponding to rows and columns.

In some cases the inflectional variation seemed to be due to a change of gender rather than of case, as could be seen from the use of these words with the ideograms for men and women. This enabled the deduction of groups of symbols which all contained the same vowel, using the following reasoning: if the masculines all form their feminines alike as in Latin (where, for example, the masculine form *dominus* becomes *domina* in the feminine), then, from the table of similar syllables, links may be deduced between symbols containing the same vowel.

Ventris deduced that symbol 08 corresponded to *a* because of its high initial frequency; this implied that consonant VIII was *n*, because the Cypriot symbol for *na* is identifiable with symbol 06, and vowel I was probably *i*, because the Cypriot symbol for *ti* is almost identical with Linear B symbol 37. Chadwick describes how further symbols were identified by searching for whole words in the tablets. For example, a name which would be expected to occur in texts written at Knossos is that of the nearby harbour town, Amnisos, which is mentioned by Homer. The consonant group -*mn*- will have to be spelled out by inserting an extra vowel, since every consonant must be followed by a vowel. It should therefore have the approximate form *a-mi-ni-so* or *08-?-?-30*. Ventris found only one suitable candidate word in the tablets. The symbol sequence for *ko-no-so*, meaning *Knossos*, was also found, since it consisted of three syllables all ending in *o*, and was *70-52-12*.

Eventually Ventris came to believe that the language underlying Linear B was in fact Greek. The word *koriannon*, which is Greek for *coriander* was found, but since this may have been a borrowed term, its presence in the tablets is not conclusive proof that Linear B is Greek. However, after the identification of *harmata*, meaning *chariots* and *a-ra-ru-ja a-ni-ja-phi*, meaning *fitted with reins*, the solution that the words in Linear B were Greek was inescapable. Blegen found further evidence that Linear B was Greek, by using Ventris's syllabary to show that the word preceding a pictograph of a tripod was *ti-ri-po-de* and the word preceding a pictograph of a four-handled pot was *qe-to-ro-we* where *owe* corresponded to the Greek word for *ear* or *handle*. Similarly, a pot with no handles was *a-no-we*, where *an* means *not*. Kahn (1966) gives examples of the content of the Linear B tablets, which record relatively minor commercial transactions, as *Koldos the shepherd holds a lease from the village, 48 litres of wheat*, (sic), and *One pair of wheels bound with bronze, unfit for service*. Before the decipherment of Linear B, the oldest known example of European writing originated from about 750 BC – the language written in Linear B was some 700 years older than this.

4.2 The Hackness cross cryptic inscriptions
The Hackness cross, originally erected in the 8th or 9th century, now consists of two damaged stone fragments which have a combined height of about 1.5 metres (Sermon 1996a, b). The fragments are now located in St Peter's church at Hackness in North Yorkshire. On the cross are five inscriptions, three in Latin and two cryptic inscriptions. The Latin inscriptions read as shown in Table 5.21.

Oedilburga, or Aethelburg, was probably abbess of the monastery at Hackness. One of the cryptic inscriptions is written in a form of Ogham, a Celtic alphabet developed in the 4th century. It consists of 27 letters forming a

---

OEDI)L(BVR)GA SEMPER TENENT MEMORES COMMV(NITATE)S TVAE TE MATER AMANTISSIMA

Oedilburga your communities hold you always in memory most loving mother

TREL(...)OSA ABATISSA OEDILBVRGA ORATE PR(O NOBIS)

Trel..osa Abbess Oedilburga pray for us

OEDILBV(RGA) BEATA A(D S)EMPER T(E REC)OLA(NT)

Blessed Oedilburga always may they remember you

Table 5.21 The Latin inscriptions on the Hackness cross

four-line inscription. In the Celtic Ogham alphabet, a fixed alphabet is divided into groups of five letters. A specific type of stroke (such as long vertical line or right sloping line) is used for each group, the letters in each group being distinguished by the number of strokes used. The groups are as follows: *B L V S N*; *H D T C Q*; *M G Ng Z R*; *A O U E I*; *Kh Th P Ph X*; *EA OI IA UI AE*. The alphabet used on the Hackness cross also consists of six groups of five letters, as shown in Table 5.22 below.

Group A	|	||	|||	||||	|||||
Group B	-	- -	- - -	- - - -	- - - - -
Group C	/	//	///	////	/////
Group D	\	\\	\\\	\\\\	\\\\\
Group E	(	((	(((	((((	(((((
Group F	)	))	)))	))))	)))))

Table 5.22 The Hackness cross alphabet, divided into six groups of five letters

Of this alphabet of 30 letters only 14 are used in the inscription. In order to decode the inscription, it is necessary to know the alphabet on which the Hackness Ogham script is based. Since it consists of 30 letters, we can exclude the Greek or Latin alphabets which have too few characters. The alternative options examined were Celtic Ogham and Anglo-Saxon Runic. The order of letters in the Anglo-Saxon Runic alphabet is *f u th o r c g w h n i j e p x s t b e m l ng oe d a ae y ea io k g q st* – these characters were divided into six groups of five.

We need to know which of the letter groups in the Celtic or Runic alphabets corresponds to which of the letter groups in the Hackness alphabet. Since there are six letter groups, there are a total of 720 different possible permutations. A computer program was written to generate all the permutations for each of the two alphabets, giving 1440 possible readings of the inscription. The most promising permutation was found using the Celtic Ogham alphabet, which produced the Old Irish interpretation shown in Table 5.23. The alphabet which produces this interpretation is shown in Table 5.24, and the most probable translation of the words on the cross is shown in Table 5.25.

- - - -	|||||	- -	((((	||			
E	R	O	S	G			

|||||	\	||	- - - -	||	- - -	((((	- -
R	H	G	E	G	U	S	O

\\\\	|||||	||	- - - -	|||	))))	////	|||||
C	R	G	E	Ng	Ph	Ui	R

////	\\\	- - - -	|||	//	||||		
Ui	T	E	Ng	Oi	Z		

Table 5.23 The Hackness Ogham inscription and Old Irish interpretation

(B	((L	(((V	((((S	(((((N
\ H	\\ D	\\\ T	\\\\ C	\\\\\ Q
| M	|| G	||| Ng	|||| Z	||||| R
- A	- - O	- - - U	- - - E	- - - - - I
) Kh	)) Th	))) P	)))) Ph	))))) X
/ Ea	// Oi	/// Ia	//// Ui	///// Ae

Table 5.24 The Hackness Ogham alphabet

Reconstruction	Old Irish	Interpretation
Ceros gu	Cross cu	Cross to
Rhge Guso	Rig Isu	King Jesus
crg eng phuir	carric an foir	rock of help
uit Engoiz	uait Oengus	from Angus

Table 5.25 Most probable interpretation of the words on the Hackness cross

An Ogham inscription in Old Irish could also have been used in North Yorkshire during the 8th and 9th centuries, considering the Celtic origins of Christianity in Northumbria.

Another cryptic inscription on the cross consists of 15 Anglo-Saxon runes, 35 'tree' runes and three Latin letters. The tree runes consist of a central vertical 'trunk' with up to four 'branches' on the left side and up to eight 'branches' on the right side. This yields an alphabet of 32 letters, consisting of four groups of eight letters, each group having a fixed number of left-side 'branches'. However, attempts to make this alphabet coincide with the letters of the Anglo-Saxon Runic alphabet were not successful.

Sermon also divided the Anglo-Saxon runic alphabet into four groups of eight letters. He then needed to find out which group of Anglo-Saxon letters corresponded to which group of tree runes. Four groups of letters can be arranged in 24 different orders, so again a computer program was used to generate all possible permutations. It was also decided to run the program for groups generated by using the runic alphabet in reverse order. This generated 48 possible readings of the inscription, none of which appeared to form any intelligible pattern. Sermon concluded that the tree runes were too fragmentary to ever be fully understood.

It was proposed that the runes of the inscription, corresponding to the sequence + *E M B D W OE G N L G U I OE R* were an anagram of *Oedilburg gnoew me* which corresponds to the Anglo-Saxon *Aethelburg cneow me* meaning *Aethelburg knew me*. The three Latin letters at the end of the tree runes were *ORA,* meaning *pray*. Sermon's work is also described by Geake (1994).

4.3 A statistical approach to machine translation

Brown et al. (1990) consider the translation of individual sentences, assuming initially that every sentence in one language is a possible translation of any sentence in the other. Every possible pair of sentences is assigned a probability called the translation probability, denoted $Pr(T \mid S)$ which is the probability that sentence T in the target language (the language we are translating into) is the correct translation of sentence S in the source language (the language we are translating from). A second probability that must be considered is the language model probability denoted $Pr(S)$. The task is to search among all possible source sentences to select the sentence S that maximises the product $Pr(S) \times Pr(T \mid S)$.

For the language model, Brown et al. suggest using an **n-gram** model, which considers the probability of n words occurring in sequence. To demonstrate the power of a **trigram** model (considering the probabilities of sequences of three words), they performed the task of **bag translation**. In bag translation, a sentence is cut up into words and then an attempt is made to reconstruct the sentence. The n-gram model is used to calculate which is the most likely of all possible arrangements of the words, by multiplying together the probabilities of all the trigrams which make up the reconstituted sentence.

For the translation model, Brown et al. take the French translation of an English sentence as being generated from the English sentence word by word. For example, in the sentences *John loves Mary* and *Jean aime Marie, John* aligns with *Jean, loves* aligns with *aime* and *Mary* aligns with *Marie*. The number of French words that an English word produces in an alignment is called its **fertility** in that alignment. For example, the English word *nobody* usually has a fertility of two since it normally produces both *ne* and *personne* in the French equivalent sentence. The term **distortion** is used whenever the word order is not preserved exactly in translation; for example, where an adjective precedes the noun it modifies in English but follows it in French. To illustrate the notation used by Brown et al. for alignments, consider the sentences *John does* *beat the dog* and *Le chien est battu par Jean.* The alignment is denoted (*Le chien est battu par Jean | John(6) does beat(3,4) the(1) dog(2)*), showing that *John* produces the sixth word in the French sentence, *does* produces nothing, *beat* produces both the third and fourth words (*est battu*) and so on. *Par* is not produced by any of the English words. To compute the probability of this alignment, the following calculation is performed:

$Pr(fertility = 1 \mid John)$ x $Pr(Jean \mid John)$ x
$Pr(fertility = 0 \mid does)$ x
$Pr(fertility = 2 \mid beat)$ x $Pr(est \mid beat)$ x $Pr(battu \mid beat)$ x
$Pr(fertility = 1 \mid the)$ x $Pr(Le \mid the)$ x
$Pr(fertility = 1 \mid dog)$ x $Pr(chien \mid dog)$ x
$Pr(fertility = 1 \mid <null>)$ x $Pr(par \mid <null>)$.

The first line of this equation means multiply the probability that *John* has a fertility of 1 by the probability that *Jean* is the translation of *John*. $Pr(par \mid <null>)$ is the probability that the word *par* is produced from nothing in the equivalent English sentence. The above equation must next be multiplied by the distortion probabilities. These are in the form $Pr(i \mid j,l)$ where i is a target position, j a source position, and l the target length. Thus, $Pr(1 \mid 6,6)$ is the probability that the first word of the target language was produced by the sixth word of the source language, if the target sentence is six words long. The parameters of the translation model are thus the set of fertility probabilities, the set of translation probabilities and the set of distortion probabilities.

To search for the sentence S that maximises the product of the language and translation models, Brown et al. start by assuming that the target sentence was produced by a sequence of source words that we do not know. For the sentence *Jean aime Marie*, this is denoted (*Jean aime Marie | **) where the asterisk denotes an unknown sequence of source words. At the first iteration, they try out all the possibilities where a single word of the source sentence is tried out at a given position, such as in the case (*Jean aime Marie | John(1)**) which is the probability of *John* producing the first word of the French sentence and the rest of the French sentence being produced by an unknown English sequence, or (*Jean aime Marie | * cat(2)**). The attempts with the highest probability of being correct are saved, and extended by trying out a second word at the next iteration. The search ends when the likeliest complete sentence alignment is found.

The parameters of the language and translation models are estimated from a large quantity of real data. In order to estimate the parameters of the language model, where a **bigram** model was employed, Brown et al. used the English-language portion of the Canadian *Hansards*, while to estimate the parameters of the translation model, pairs of sentences that are mutual translations (such as found by comparing both the English and French sections of the Canadian *Hansards*) were required.

To find out which sentences in the English portion of the *Hansards* corresponded to which sentences in the French portion, Brown et al. used a statistical algorithm based on sentence length. However, the resulting sentence pairs were still not aligned at the word level. Thus, it was not possible to estimate the translation model parameters by simple counting. Instead the EM algorithm (described in Chapter 2, Section 2.10.3) was used by Brown et al. to estimate the parameters of the translation model, as described below:

> Given some initial estimate of the parameters, we can compute the probability of any particular alignment. We can then re-estimate the parameters by weighing each possible alignment according to its probability as determined by the initial guess of the parameters. Repeated iterations of this process lead to parameters that assign ever greater probability to the set of sentence pairs that we actually observe. (1990, p. 82)

In their pilot experiment, the translation and fertility probabilities for the English word *not* were found as shown in Table 5.26.

French	Probability	Fertility	Probability
pas	0.469	2	0.758
ne	0.460	0	0.133
non	0.024	1	0.106
pas du tout	0.003		
faux	0.003		
plus	0.002		
ce	0.002		
que	0.002		
jamais	0.002		

Table 5.26 Translation and fertility probabilities for *not*

Not surprisingly, *pas* appears as the most probable translation. The fertility probabilities show that *not* most often translates into two words *ne ... pas*. However, when these probabilities were used to attempt the translation of French into English, fewer than half the translations were acceptable.

Sections 4.1 to 4.3 have illustrated how statistical techniques are not only useful in attributing the authorship or determining the chronology of text, but given a corpus of data which is in some sense encoded (such as being written in an unknown script or foreign language), we can use statistical techniques to analyse that corpus and produce significant results: the translation of known languages or even the decipherment of an unknown language.

5 SUMMARY

The themes of literary detective work that we have explored in this chapter are:

- computer studies of stylometry, including studies of disputed authorship
- language relationship and divergence
- translation.

Criteria which have been successfully employed in determining authorship in cases of dispute are

- usage of function words
- word placement within a sentence
- proportional pairs or near-synonyms, one of which is favoured more by one author than another
- sentence length
- Yule's K characteristic, a measure of the probability that any randomly selected pair of words will be identical
- vocabulary unique to a particular sample
- Bayesian statistics, where a distribution such as the Poisson distribution is used to estimate the probabilities A and B of a word occurring a given number of times in two samples of text of known authorship. Once the actual number of occurrences of that word in a sample of disputed text is known, the overall probability of that text being written by one of the authors is updated by multiplying the prior probability by the ratio of A and B
- Milic's D measure, the total number of different part of speech trigrams used, and
- the chi-square test to compare the relative usage of different parts of speech in two texts.

Related to the question of disputed authorship is the question of an author's writing style changing over time. This may be a smooth progressive change over a lifetime, or may vary according to the choice of pseudonym adopted, or to circumstance, where a political writer can be more dogmatic in times of crisis and more tentative in times of success. Similarly related to the theme of disputed authorship is the field of forensic stylometry, where distinctions are made between real and fabricated confessions, and the authorship of anonymous letters, contracts and wills is established. In each case authentic textual material to represent the accused must be provided.

In studies of linguistic relationship, we look for the presence of cognate terms which remain in the daughter languages having originated in the antecedent language. A contingency table can be drawn up to record the number of words in the antecedent language which remain in both, one or neither daughter languages. Different formulae for the correlation coefficient between the daughter languages based on the contingency table apply depending on whether the vocabulary of the ancestor language is known or unknown. To examine the question of language divergence, we must estimate the retention rate of words in a language. By analogy with the process of radioactive decay, the disappearance of an individual cognate term from a language is a random process, but the overall pattern of decline in the number of cognate terms in a language as a function of time may be predicted. Different retention rates are found in words of different meanings.

In the translation of Linear B, neither the language nor the script was originally known. The size of the alphabet suggested that the script was probably syllabic. Grouping of similar words into three cases revealed the common inflectional endings, and the existence of 'bridges' which were syllables consisting of the last letter of the root word and the first letter of the ending. Using this data, a grid could be composed showing the syllables which had consonants and vowels in common. The identification of words such as place names helped show that the underlying language was Greek. To decipher the Ogham-like script on the Hackness cross, where the alphabet consisted of six clear groups of five characters, known contemporary alphabets were also divided into groups of five contiguous letters. All possible permutations of these groups were generated by computer, to see which produced pronounceable readings of the runes on the cross. One permutation was decided upon because the resulting transcription corresponded with Old Irish. The statistical translation procedure of Brown et al. (1990) consists of a translation model which suggests words from the source language that might have produced the words observed in the target sentence and a language model which suggests an order in which to place those source words.

6 EXERCISES

1. The author John Lancaster uses the word *while* twice every 10,000 words on average. A second author, Richard York, uses the word *while* four times every 10,000 words on average. An anonymous work of length 10,000 words is discovered, which contains three occurrences of *while*. Use the Poisson distribution to estimate who is the more likely author of this anonymous work, Lancaster or York, and the odds in that author's favour. Assume that the prior odds, based on historical evidence, for Lancaster being the author as opposed to York are 1 to 1.

2. What is Yule's K Characteristic for a text consisting only of the words *Home Sweet Home*?

3. A sentence of a book is part of speech tagged as follows:

   this_ART book_NOUN will_VERB finish_VERB with_PREP a_ART detailed_ADJ examination_NOUN of_PREP literary_ADJ detective_ADJ work_NOUN

   What is Milic's D value for bigrams (pairs of adjacent tags)?

4. Sundanese, Javanese, Malay and Madurese are all daughter languages of the no longer spoken Proto-Malayo-Javanic. The words for *animal*, *bird*, *dog*, *fish*, *louse* and *snake* in each of the four daughter languages (Nothofer 1975) are presented below:

	Sundanese	Javanese	Malay	Madurese
animal	binatang	kewan	binatang	bhurun alas
bird	manuk	mano'	burung	mano'
dog	'anjing	asu	anjing	pate'
fish	lauk	iwaq	ikan	jhuko'
louse	kutuk	lingso	kutu	koto
snake	'oray	ulo	ular	olar

Considering only those words for which the original Proto-Malayo-Javanic can be estimated, find the correlation coefficient r_n for the language pair Malay and Madurese.

7 FURTHER READING

For a broader-based account of author identification, which includes such topics as historical evidence and multivariate statistics, consult *Authorship Identification and Computational Stylometry*, by McEnery and Oakes (1997). Another useful review of recent developments in automated stylometry is given by Holmes (1994). Chadwick's (1958) account of the decipherment of Linear B is highly accessible to a non-specialist readership.

NOTE

1. The Nijmegen corpus consists of texts from a wide range of subjects which have been annotated with two different syntactic analysis systems. Only the crime fiction texts from the corpus have been used here.

Glossary

a posteriori probability the probability of a hypothesis after new evidence becomes available.

a priori probability the probability of a hypothesis before new evidence becomes available.

abduction reasoning from evidence to hypothesis.

absolute magnitude for a positive value, the value itself; for a negative value, the value when the minus sign has been replaced by a plus sign.

agglomerative clustering to start with a number of small clusters, then sequentially merge them until one large cluster is left.

algorithm a way of performing a particular task, often incorporated into a computer program.

alignment the practice of defining explicit links between texts in a parallel corpus.

alignment distance the number of operations such as insertion, deletion or substitution required to transform one sequence into another.

annotation (i) the practice of adding explicit additional information to machine-readable text; (ii) the physical representation of such information.

approximate string matching the identification of words with similar character sequences for error correction or information retrieval.

association the relationship formed when two variables such as animacy of a noun and use of the genitive are related, so that the presence or level of one variable makes a difference to the distribution of observations on the other.

association criteria to compute the strength of a bond between the two lemmas of a pair, enabling lemma pairs to be sorted from the most tightly to the least tightly bound.

asymptotic a line which is asymptotic to another line becomes increasingly closer to it without ever touching it.

base see **logarithm**.

Baum–Welch algorithm one of a class of algorithms called E-M (estimation-maximisation) which adjusts the parameters (such as the transition probabilities) of a **hidden Markov model** to optimise its performance.

Bayesian statistics a branch of statistics where we talk about belief in a hypothesis rather than its absolute probability; this degree of belief may change given new evidence.

bigram a sequence of two consecutive items such as letters or words.

binomial coefficient the number of ways of selecting r objects out of n; for example, when selecting two letters out of three (ABC), we can take (AB), (AC) or (BC), yielding three possibilities.

binomial distribution if n independent experiments are performed, each with the same probability of success (for example, an unbiased coin is tossed five times, where a head is deemed a success), the probability of obtaining each possible number of successes (from 0 to 5 in our example) is shown by the binomial distribution.

binomial probability graph a graph of the **binomial distribution**.

block sampling a method of sampling where the starting point of the sample is randomly chosen; the sample is then a single continuous portion of text of desired length beginning at the starting point.

categorisation the initial identification of classes for classification, which must take place before classification.

central limit theorem when samples are repeatedly drawn from a population, the means of the samples will be normally distributed around the population mean.

character a single letter, number, punctuation mark or other symbol.

class exemplar see **cluster centroid**.

classification the assignment of objects to predefined classes.

cloze procedure a test in which certain words in a text are blanked out and a subject has to guess the missing words.

cluster analysis the discovery of a category structure; finding the natural groups in a series of observations such as texts or words.

cluster centroid clusters can be represented by their centroid, which is in some way the 'average' of all the cluster members, sometimes called a **class exemplar**.

clustering the grouping of similar objects, and the keeping apart of dissimilar objects.

coefficient in the expression involving various powers of x such as $ax^2 + bx$, a is the coefficient of x^2 and b is the coefficient of x.

collocation the patterns of combinations of words (for example, with other words) in a text.

common factor variance the amount of variance shared by two variables in **principal components analysis**.

communality of a variable in **principal components analysis** is the sum of all the common factor variance of the variable over all factors, which is the variance it shares in common with the other variables.

complete linkage or **furthest neighbour clustering** differs from single linkage in that the similarity between clusters is calculated on the basis of the least similar pair of documents, one from each cluster.

component loadings show how the original variables correlate with the principal component in **principal components analysis**.

concordance comprehensive listing of a given item in a corpus (most often a word or a phrase), also showing its immediate context.

contingency table method of presenting the frequencies of the outcomes from an experiment in which the observations in the sample are classified according to two criteria (Clapham 1996).

continuous data variables which can take any value, not being constrained to a number of discrete values.

cophenetic correlation coefficient the most common distortion measure, produced by comparing the values in the original similarity matrix with the interobject similarities found in the resulting dendrogram.

corpus (i) (loosely) any body of text; (ii) (most commonly) a body of machine-readable text; (iii) (more strictly) a finite collection of machine-readable text, sampled to be maximally representative of a language or variety.

curvilinear relations variables are related in curvilinear fashion if a graph of one plotted against the other produces a curve.

deduction reasoning from hypothesis to evidence.

dendrogram a tree diagram where the branch points show the similarity level at which two entities (data items or clusters) fuse to form a single cluster.

dependent variable variable thought to be influenced by one or more independent variables.

dispersion measures show, for example, whether linguistic features are evenly spread throughout the corpus or whether they tend to clump together.

dissimilarity measures measures for comparing pairs of items such as sets of index terms or character strings; a high score indicates a low degree of similarity.

distortion measures show how faithfully or otherwise a set of clusters represents the original data.

divisive clustering to start with one large cluster containing the entire data set, then sequentially divide and subdivide it until we are left with many small clusters, perhaps containing one data item each.

e mathematical constant, equal to about 2.71828183; *e* to the power *x* is called the exponential of *x*, denoted exp *x*.

eigenvalues the amount of variance accounted for by each component in **principal components analysis**.

elastic matching see **time warping**.

E-M algorithm see **Baum–Welch algorithm**.

e-mail communication system which sends messages and data very quickly to computer sites around the world.

entropy measure of randomness in nature; the amount of information in a message is formally measured by the entropy of the message.

equiprobable equally likely.

equivalence if two character strings which are superficially different can be substituted for each other in all contexts without making any difference of meaning, then they are said to be equivalent.

ergodic Markov models where every state of the model can be reached from every other state in a finite number of steps.

extracted variance in **principal components analysis**, the sum of the squares of the loadings of the variables on a factor.

factor analysis multivariate statistical procedure used to reduce the apparent dimensionality of the data; does not lead to a unique solution, so **factor rotation** is required.

factor loading in **principal components analysis** the correlation between a variable and a factor is called the loading of the variable on the factor.

factor rotation allows an experimenter to choose a preferred solution to **factor analysis** from an infinite set of possible solutions.

finite state automata consist of nodes representing states and branches representing transitions from one state to another; a transition from state A to state B takes place when the symbol attached to branch AB occurs at the input (Salton and McGill 1983).

flexible collocation collocation in which the words may be inflected, the word order may vary and the words can be separated by any number of intervening words.

forward–backward algorithm reduces the number of calculations required to evaluate a **hidden Markov model**, that is, compute the probability that a particular sequence of observed outputs would occur as a result of this model.

frequency list list of linguistic units of a text (often words) which also shows their frequency of occurrence within a text.

frequency polygon to produce a frequency polygon, plot frequency (the number of data items with a given value) on the y-axis against each possible data value on the x-axis; then draw straight lines to connect adjacent points on the graph.

furthest neighbour clustering see **complete linkage**.

hapax legomena all the vocabulary items in a text which occur just once each.

hidden Markov model instead of emitting the same symbol each time at a given state (which is the case for the **observable Markov model**), there is now a choice of symbols, each with a certain probability of being selected.

independent variable the variable consciously varied by the experimenter, as opposed to the **dependent variable** which is merely observed in response to changes in the independent variable.

information the amount of information in a message is the average number of bits needed to encode all possible messages in an optimal encoding; for example, the sex field in a database contains one bit of information because it can be coded with one bit if *male* is replaced by 0 and *female* is replaced by 1 (Denning 1982).

interaction when three variables are related, in such a way that the association between two of them (such as animacy of a noun and the use of the genitive) changes according to the nature or level of the third variable (for example, genre), the relationship between the three is called interaction.

interpolation method of estimating a value which is known to fall between two other values.

interval scale as for **ratio scale** except that the zero point is arbitrary, for example, temperature in degrees centigrade.

inverted file for each term in the lexicon, an associated list of line, paragraph or document reference numbers is given; each reference uniquely specifies the location to which a given term has been assigned.

keyword word input to a concordance program to obtain the required lines of text.

KWAL keyword and line, a form of concordance which can allow several lines of context either side of the keyword.

KWIC keyword in context, a form of concordance in which a word is given within x words of context either side of the keyword.

lemma headword form that one would look for if looking up a word in a dictionary, for example, the word-form *loves* belongs to the lemma *love*.

lexicon essentially synonymous with *dictionary* – a collection of words and information about them; this term is used more commonly than *dictionary* for machine-readable dictionary databases.

linear describing a straight line.

linguistics the science of language.

logarithm $\log_a x$ is called the logarithm to the base a of x; this is the power to which a must be raised in order to get x; for example, $\log_2 (8) = 3$ because $8 = 2 \times 2 \times 2$.

logistic regression type of loglinear analysis where there is one dependent variable which is to be explained by other independent variables.

logit function function of the probability of an event; the natural logarithm of the ratio of the probability of the event occurring to the probability of that event not occurring, given by the formula $f(p) = \log_e (p/(1-p))$.

loglinear analysis means of modelling tabular data involving three or more variables.

lower-order relatives marginal tables that can be derived from a particular marginal table are known as its lower-order relatives.

marginal consider a 2 x 2 matrix, with values a and b in the top row and c and d in the bottom row. The row marginals will be $a+b$ and $c+d$, and the column marginals will be $a+c$ and $b+d$.

Markov model discrete **stochastic** process, where the probability of each possible state being reached at the next time interval depends only on the current state and not on any previous states.

matrix rectangular array of values displayed in rows and columns. An $m \times n$ matrix has m rows and n columns.

mean average of all values in a data set, found by adding up all of the values then dividing by the number of values.

measure of central tendency most typical score for a data set; the three common measures of central tendency are the **mode, median** and **mean.**

median central value in a data set, where half the values will be above the median and the other half below the median.

mode most frequently obtained value in a data set.

monothetic classification employing the Aristotelean definition of a class; all class members must have a certain set of properties for membership in the class.

monotonic function a function of x is monotonic if, whenever x increases, the function either increases or stays the same, and whenever x decreases, the function either decreases or stays the same.

multidimensional scaling technique which constructs a pictorial representation of the relationships inherent in a dissimilarity matrix.

multinomial distribution related to **binomial distribution,** except that more than two outcomes are possible for each experiment.

multiple regression regression is multiple if there are two or more independent variables.

multivariate statistics those statistical methods which deal with the relationships between many different variables.

mutual information probability of two things happening together compared with the probability of their occurring independently; it is thus a statistical measure of the degree of relatedness of two elements.

Napierian logarithms see **natural logarithms.**

natural logarithms logarithms to the base e are called natural or **Napierian logarithms,** and the notation ln can be used instead of $\log_e$ (Clapham 1996).

nearest neighbour clustering see **single linkage.**

n-gram a sequence of n consecutive items such as letters or words.

node in collocation analysis, the word whose collocates are being investigated; **pole**.

nominal data the observations are not numeric or quantitative, but are descriptive and have no natural order (Clapham 1996).

nominal scale whenever items can be categorised quantitatively (for example, noun, verb, adjective) but the numbers we assign to the categories are arbitrary and do not reflect the primacy of any one category over the others.

non-continuous variables can only take one of a number of discrete values.

non-metric scaling one of a number of techniques for mapping subjectively judged similarity data.

non-parametric tests statistical tests that make no assumptions about the underlying population distribution. Such tests often use the median of a population and the rank order of the observations.

null hypothesis denoted H_0, a particular assertion that is to be accepted or rejected. To decide whether H_0 is to be accepted or rejected, a significance test tests whether a sample taken from a population could have occurred by chance, given that H_0 is true (Clapham 1996).

observable Markov model see **hidden Markov model**.

optimal codes coding systems which are free of redundancy.

ordinal scale where the order of items rather than the difference between them is measured.

orthogonal two lines which are orthogonal are at right angles to each other.

parallel corpus a corpus which contains the same texts in more than one language.

parametric tests statistical tests which assume that the underlying population is normally distributed, and the mean and standard deviation are appropriate measures of central tendency and dispersion.

perplexity the perplexity of a message is the size of the set of equiprobable events which has the same information.

Poisson distribution gives the number of occurrences in a given time of an event which occurs randomly but at a given rate (Clapham 1996).

pole see **node**.

polythetic categorisation items such as documents are placed in the cluster that has the greatest number of attributes such as index terms in common, but there is no single attribute which is a prerequisite for cluster membership.

population body of data about which hypotheses are drawn, based on a sample taken from that body of data.

power (of a number) a to the power b, or a^b, means that b instances of a are multiplied together, for example, $5^3 = 5 \times 5 \times 5$.

power (of a statistical test) the probability that the test rejects the null hypothesis when it is indeed false (Clapham 1996).

principal components analysis multivariate statistical procedure used to reduce the apparent dimensionality of the data; leads to a unique solution.

quantise to limit a continuous variable such as the amplitude of a speech waveform to values that are integral multiples of a basic unit.

random (i) having a value which cannot be determined but only described probabilistically; for example, we cannot say beforehand what the outcome of a throw of a dice will be, but we can say that there is a one in six chance it will be a six; (ii) chosen without regard to any characteristics of the individual members of the population, so each has an equal chance of being selected (Hanks 1986).

range highest value in a data set minus the lowest value.

rank order data is said to be in rank order or ranked when arranged in ascending or decending order of magnitude according to some observable feature; for example, the text with the greatest number of words will be ranked first according to size.

ratio scale exemplified by measurement in centimetres; each unit on the scale is the same as each other unit, and thus the difference between 1cm and 2cm is the same as the difference between 9cm and 10cm.

redundancy a measure of how the length of text is increased due to the statistical and linguistic rules governing a language.

regression statistical procedure to determine the relationship between a dependent variable and one or more independent variables (Clapham 1996).

regression line line which runs most closely through the points of a **scatter plot**.

reification to interpret meaningfully each component found in **principal components analysis**.

relative entropy ratio of actual entropy divided by the maximum entropy.

saturation model in loglinear analysis, a model in which all the variables interact and which fits the data perfectly.

scatter plot or **scatter diagram** two dimensional diagram showing values of an **independent variable** plotted against the corresponding values of the **dependent variable**.

seed point first member of a new cluster.

Shannon diversity depends both on the frequency with which a **lemma** is found in a lemma pair and the number of different lemma pairs in which it is found.

significance/significant reaching a degree of statistical certainty at which it is unlikely that a result is due purely to chance.

similarity degree to which two character strings resemble each other; unlike equivalence, this is not an all-or-nothing phenomenon, but quantified by a real valued metric.

simple regression regression is simple if there is only one **independent variable**.

single linkage the best known of the agglomerative clustering methods; clusters are joined at each stage by the single shortest or strongest link between them; also referred to as **nearest neighbour clustering**.

spell checker in word processing, a tool which checks that words have been spelt correctly.

spread sampling as **block sampling**, but requiring the selection of much smaller samples at different starting points.

standard deviation measure which takes into account the distance of every data item from the **mean**; the square root of **variance**.

standard error of estimate when taking a number of samples from a population, the variability of the sample means is estimated by the standard error of the **mean**.

standard inter-quartile range the difference between the value one quarter of the way from the top end of a distribution and the value three quarters of the way down in that distribution.

statistics (i) any systematic collection and tabulation of meaningful, related facts and data; (ii) the systematic study and interpretation of such a collection of facts and data (Pei and Gaynor 1954).

stemming rules rules for the removal and replacement of common prefixes and suffixes, designed to render alternative grammatical forms of a word equivalent.

stochastic matrix all the entries are non-negative; in a row–stochastic matrix the entries in each row add up to 1; a two-dimensional transition matrix where the starting states are represented by rows and the final states are represented by columns is row-stochastic.

stochastic process family of random variables; the possible values taken by the random variables are called states, and these form the state space (Clapham 1996).

stoplist negative dictionary; a list of words not to be considered in a text processing operation.

stratified random sampling sampling technique where different sections of the overall population (such as age groups for people) can be represented in the same proportion as they occur in the population as a whole, but within each section the members of the sample are chosen by random methods.

string in computer processing, a sequence of characters.

sublanguage constrained variety of a language. Although a sublanguage may be naturally occurring, its key feature is that it lacks the productivity generally associated with language.

t **test** statistical significance test based on the difference between observed and expected results.

tag a code attached to words in a text representing some feature or set of features relating to those words, such as grammatical part of speech.

tagging marking items in a text with additional information, often relating to their linguistic properties.

tagset collection of tags in the form of a scheme for annotating corpora.

term name, expression or word used for some particular thing, especially in a specialised field of knowledge, for example, a medical term (McLeod 1987).

thesaurus lexicographic work which arranges words in meaning-related groups rather than alphabetically.

time warping or **elastic matching** the need to expand or compress the time axis of a speech signal at various times, when the rate of speaking increases and decreases from instant to instant.

tokens in a word frequency list, individual occurrences of words; the number of tokens in a text is the same as the total number of words.

transition matrix a repository of transitional probabilities. This is an n-dimensional matrix, depending on the length of transition under question. So, for example, with a bigram transition, we require a two-dimensional matrix.

transition probabilities in a **Markov model**, the probability that the state will be j, given that the previous state was i. The transition probability is denoted p_{ij}.

trigram sequence of three consecutive items such as letters or words.

truncation with simple truncation, an equivalence class consists of all terms beginning with the same n characters.

types number of types in a text is the number of unique word forms, rather than the total number of words in the text.

unicity distance number of letters of a coded message needed to achieve a unique and unambiguous solution when the original message (or plain text) has a known degree of redundancy.

univariate concerned with the distribution of a single variable.

variable rule analysis form of statistical analysis which tests the effects of combinations of different factors and attempts to show which combination accounts best for the data being analysed.

variance measure which takes into account the distance of every data item from the mean; equal to **standard deviation** squared.

Viterbi algorithm reduces the number of calculations required to estimate the most likely sequence of states that a **hidden Markov model** passed through, given the observed sequence of outputs from that model.

word index see **inverted file**.

z score statistical measure of the closeness of an element to the mean value for all the elements in a group.

Zipf's law according to Zipf's law, the rank of a word in a word frequency list ordered by descending frequency of occurrence is inversely related to its frequency, as shown by the formula *frequency* = k x *rank*$^{-\gamma}$, where k and γ are empirically found constants.

Appendices

APPENDIX 1   THE NORMAL DISTRIBUTION

The table gives the proportion of the total area under the curve which lies beyond any given z value (that is, the shaded area in the diagram). It is therefore appropriate for a one-tailed (directional) test. For a two-tailed (non-directional) test, the proportions must be doubled.

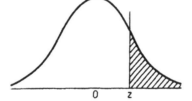

The figures down the left-hand side give values of z to the first decimal place, and those across the top give the second decimal place.

z	0.00	0.01	0.02	0.03	0.04	0.05	0.06	0.07	0.08	0.09
0.0	0.5000	0.4960	0.4920	0.4880	0.4840	0.4801	0.4761	0.4721	0.4681	0.4641
0.1	0.4602	0.4562	0.4522	0.4483	0.4443	0.4404	0.4364	0.4325	0.4286	0.4247
0.2	0.4207	0.4168	0.4129	0.4090	0.4052	0.4013	0.3974	0.3936	0.3897	0.3859
0.3	0.3821	0.3783	0.3745	0.3707	0.3669	0.3632	0.3594	0.3557	0.3520	0.3483
0.4	0.3446	0.3409	0.3372	0.3336	0.3300	0.3264	0.3228	0.3192	0.3156	0.3121
0.5	0.3085	0.3050	0.3015	0.2981	0.2946	0.2912	0.2877	0.2843	0.2810	0.2776
0.6	0.2743	0.2709	0.2676	0.2643	0.2611	0.2578	0.2546	0.2514	0.2483	0.2451
0.7	0.2420	0.2389	0.2358	0.2327	0.2296	0.2266	0.2236	0.2206	0.2177	0.2148
0.8	0.2119	0.2090	0.2061	0.2033	0.2005	0.1977	0.1949	0.1922	0.1894	0.1867
0.9	0.1841	0.1814	0.1788	0.1762	0.1736	0.1711	0.1685	0.1660	0.1635	0.1611
1.0	0.1587	0.1562	0.1539	0.1515	0.1492	0.1469	0.1446	0.1423	0.1401	0.1379
1.1	0.1357	0.1335	0.1314	0.1292	0.1271	0.1251	0.1230	0.1210	0.1190	0.1170
1.2	0.1151	0.1131	0.1112	0.1093	0.1075	0.1056	0.1038	0.1020	0.1003	0.0985
1.3	0.0968	0.0951	0.0534	0.0918	0.0901	0.0885	0.0869	0.0853	0.0838	0.0823
1.4	0.0808	0.0793	0.0778	0.0764	0.0749	0.0735	0.0721	0.0708	0.0694	0.0681
1.5	0.0668	0.0655	0.0643	0.0630	0.0618	0.0606	0.0594	0.0582	0.0571	0.0559
1.6	0.0548	0.0537	0.0526	0.0516	0.0505	0.0495	0.0485	0.0475	0.0465	0.0455
1.7	0.0446	0.0436	0.0427	0.0418	0.0409	0.0401	0.0392	0.0384	0.0375	0.0367
1.8	0.0359	0.0351	0.0344	0.0336	0.0329	0.0322	0.0314	0.0307	0.0301	0.0294
1.9	0.0287	0.0281	0.0274	0.0268	0.0262	0.0256	0.0250	0.0244	0.0239	0.0233

z	0.00	0.01	0.02	0.03	0.04	0.05	0.06	0.07	0.08	0.09
2.0	0.0228	0.0222	0.0217	0.0212	0.0207	0.0202	0.0197	0.0192	0.0188	0.0183
2.1	0.0179	0.0174	0.0170	0.0166	0.0162	0.0158	0.0154	0.0150	0.0146	0.0143
2.2	0.0139	0.0136	0.0132	0.0129	0.0125	0.0122	0.0119	0.0116	0.0113	0.0110
2.3	0.0107	0.0104	0.0102	0.0099	0.0096	0.0094	0.0091	0.0089	0.0087	0.0084
2.4	0.0082	0.0080	0.0078	0.0075	0.0073	0.0071	0.0069	0.0068	0.0066	0.0064
2.5	0.0062	0.0060	0.0059	0.0057	0.0055	0.0054	0.0052	0.0051	0.0049	0.0048
2.6	0.0047	0.0045	0.0044	0.0043	0.0041	0.0040	0.0039	0.0038	0.0037	0.0036
2.7	0.0035	0.0034	0.0033	0.0032	0.0031	0.0030	0.0029	0.0028	0.0027	0.0026
2.8	0.0026	0.0025	0.0024	0.0023	0.0023	0.0022	0.0021	0.0021	0.0020	0.0019
2.9	0.0019	0.0018	0.0018	0.0017	0.0016	0.0016	0.0015	0.0015	0.0014	0.0014
3.0	0.0013	0.0013	0.0013	0.0012	0.0012	0.0011	0.0011	0.0011	0.0010	0.0010
3.1	0.0010	0.0009	0.0009	0.0009	0.0008	0.0008	0.0008	0.0008	0.0007	0.0007
3.2	0.0007	0.0007	0.0006	0.0006	0.0006	0.0006	0.0006	0.0005	0.0005	0.0005
3.3	0.0005	0.0005	0.0005	0.0004	0.0004	0.0004	0.0004	0.0004	0.0004	0.0003
3.4	0.0003	0.0003	0.0003	0.0003	0.0003	0.0003	0.0003	0.0003	0.0003	0.0002
3.5	0.0002	0.0002	0.0002	0.0002	0.0002	0.0002	0.0002	0.0002	0.0002	0.0002

APPENDIX 2   THE DISTRIBUTION

The table gives critical values of t for significance at various levels, in a two-tailed/non-directional or a one-tailed/directional test, for different numbers of degrees of freedom. These critical values are the values beyond which lies that proportion of the area under the curve which corresponds to the significance level.

			Significance level: two-tailed/non-directional		
	0.20	0.10	0.05	0.02	0.01
			Significance level: one-tailed/directional		
Degrees of freedom	0.10	0.05	0.025	0.01	0.005
1	3.078	6.314	12.71	31.82	63.66
2	1.886	2.920	4.303	6.965	9.925
3	1.638	2.353	3.182	4.541	5.841
4	1.533	2.132	2.776	3.747	4.604
5	1.476	2.015	2.571	3.365	4.032
6	1.440	1.943	2.447	3.143	3.707
7	1.415	1.895	2.365	2.998	3.499
8	1.397	1.860	2.306	2.896	3.355
9	1.383	1.833	2.262	2.821	3.250
10	1.372	1.812	2.228	2.764	3.169
11	1.363	1.796	2.201	2.718	3.106
12	1.356	1.782	2.179	2.681	3.055
13	1.350	1.771	2.160	2.650	3.012
14	1.345	1.761	2.145	2.624	2.977
15	1.341	1.753	2.131	2.602	2.947
16	1.337	1.746	2.120	2.583	2.921
17	1.333	1.740	2.110	2.567	2.898
18	1.330	1.734	2.101	2.552	2.878
19	1.328	1.729	2.093	2.539	2.861
20	1.325	1.725	2.086	2.528	2.845
21	1.323	1.721	2.080	2.518	2.831
22	1.321	1.717	2.074	2.508	2.819
23	1.319	1.714	2.069	2.500	2.807
24	1.318	1.711	2.064	2.492	2.797
25	1.316	1.708	2.060	2.485	2.787
26	1.315	1.706	2.056	2.479	2.779
27	1.314	1.703	2.052	2.473	2.771
28	1.313	1.701	2.048	2.467	2.763
29	1.311	1.699	2.045	2.462	2.756
30	1.310	1.697	2.042	2.457	2.750
40	1.303	1.684	2.021	2.423	2.704
60	1.296	1.671	2.000	2.390	2.660
120	1.289	1.658	1.980	2.358	2.617
∞	1.282	1.645	1.960	2.326	2.576

APPENDIX 3   THE MANN–WHITNEY U TEST

The first table gives the critical values for significance at the $p \leq 0.05$ level in a two-tailed/non-directional test, and for the $p \leq 0.025$ level in a one-tailed/directional test. The second table gives the critical values for the $p \leq 0.01$ level in a two-tailed/non-directional test, and for the $p \leq 0.0005$ level in a one-tailed/directional test. For significance, the calculated value of U most be *smaller than or equal to* the critical value. N_1 and N_2 are the number of observations in the smaller and larger group, respectively.

N_1	5	6	7	8	9	10	11	12	13	14	15	16	17	18	19	20
$p \leq 0.05$ (two-tailed), $p \leq 0.025$ (one-tailed)																
5	2	3	5	6	7	8	9	11	12	13	14	15	17	18	19	20
6		5	6	8	10	11	13	14	16	17	19	21	22	24	25	27
7			8	10	12	14	16	18	20	22	24	26	28	30	32	34
8				13	15	17	19	22	24	26	29	31	34	36	38	41
9					17	20	23	26	28	31	34	37	39	42	45	48
10						23	26	29	33	36	39	42	45	48	52	55
11							30	33	37	40	44	47	51	55	58	62
12								37	41	45	49	53	57	61	65	69
13									45	50	54	59	63	67	72	76
14										55	59	64	69	74	78	83
15											64	70	75	80	85	90
16												75	81	86	92	98
17													87	93	99	105
18														99	106	112
19															113	119
20																127
$p \leq 0.01$ (two tailed), $p \geq 0.005$ (one-tailed)																
5	0	1	1	2	3	4	5	6	7	7	8	9	10	11	12	13
6		2	3	4	5	6	7	9	10	11	12	13	15	16	17	18
7			4	6	7	9	10	12	13	15	16	18	19	21	22	24
8				7	9	11	13	15	17	18	20	22	24	26	28	30
9					11	13	16	18	20	22	24	27	29	31	33	36
10						16	18	21	24	26	29	31	34	37	39	42
11							21	24	27	30	33	36	39	42	45	48
12								27	31	34	37	41	44	47	51	54
13									34	38	42	45	49	53	57	60
14										42	46	50	54	58	63	67
15											51	55	60	64	69	73
16												60	65	70	74	79
17													70	75	81	86
18														81	87	92
19															93	99
20																105

APPENDIX 4   THE SIGN TEST

The table gives critical values of x (the number of cases with the less frequent sign) for different values of N (the number of non-tied pairs of scores). For significance, the computed value of x must be *smaller than or equal to* the critical value.

	Significance level: two-tailed/non-directional		
	0.10	*0.05*	*0.02*
	Significance level: one-tailed/directional		
N	*0.05*	*0.025*	*0.01*
5	0	–	–
6	0	0	–
7	0	0	0
8	1	0	0
9	1	1	0
10	1	1	0
11	2	1	1
12	2	2	1
13	3	2	1
14	3	2	2
15	3	3	2
16	4	3	2
17	4	4	3
18	5	4	3
19	5	4	4
20	5	5	4
21	6	5	4
22	6	5	5
23	7	6	5
24	7	6	5
25	7	7	6

APPENDIX 5   THE WILCOXON SIGNED RANKS TEST

The table gives critical values of W for different values of N (the number of non-tied pairs of scores). For significance, the calculated value must be *smaller than or equal to* the critical value.

| | Significance level: two-tailed/non-directional | |
| | 0.05 | 0.01 |
N	Significance level: one-tailed/directional 0.025	0.005
6	0	–
7	2	–
8	3	0
9	5	1
10	8	3
11	10	5
12	13	7
13	17	9
14	21	12
15	25	15
16	29	19
17	34	23
18	40	27
19	46	32
20	52	37
21	58	42
22	65	48
23	73	54
24	81	61
25	89	68

APPENDIX 6   THE *F* DISTRIBUTION

The table gives the critical values of *F* for different numbers of degrees of freedom (df) in the numerator and in the denominator of the expression for *F*. For each entry, two values are given. The upper value is the critical value for the $p \leq 0.05$ level in a one-tailed/directional test, and for the $p \leq 0.10$ level in a two-tailed/non-directional test. The lower value is the critical value for the $p \leq 0.01$ level in a one-tailed/directional test and for the $p \leq 0.02$ level in a two-tailed/non-directional test.

Df in denominator	1	2	3	4	5	6	7	8	9	10	12	15	20	30	50	∞
1	161	200	216	225	230	234	237	239	241	242	244	246	248	250	252	254
	4052	5000	5403	5625	5764	5859	5928	5981	6022	6056	6106	6157	6209	6261	6303	6366
2	18.5	19.0	19.2	19.2	19.3	19.3	19.4	19.4	19.4	19.4	19.4	19.4	19.4	19.5	19.5	19.5
	98.5	99.0	99.2	99.2	99.3	99.3	99.4	99.4	99.4	99.4	99.4	99.4	99.4	99.5	99.5	99.5
3	10.1	9.55	9.28	9.12	9.01	8.94	8.89	8.85	8.81	8.79	8.74	8.70	8.66	8.62	8.58	8.53
	34.1	30.8	29.5	28.7	28.2	27.9	27.7	27.5	27.3	27.2	27.1	26.9	26.7	26.5	26.4	26.1
4	7.71	6.94	6.59	6.39	6.26	6.16	6.09	6.04	6.00	5.96	5.91	5.86	5.80	5.75	5.70	5.63
	21.2	18.0	16.7	16.0	15.5	15.2	15.0	14.8	14.7	14.5	14.4	14.2	14.0	13.8	13.7	13.5
5	6.61	5.79	5.41	5.19	5.05	4.95	4.88	4.82	4.77	4.74	4.68	4.62	4.56	4.50	4.44	4.36
	16.3	13.3	12.1	11.4	11.0	10.7	10.5	10.3	10.2	10.1	9.89	9.72	9.55	9.38	9.24	9.02
6	5.99	5.14	4.76	4.53	4.39	4.28	4.21	4.15	4.10	4.06	4.00	3.94	3.87	3.81	3.75	3.67
	13.7	10.9	9.78	9.15	8.75	8.47	8.26	8.10	7.98	7.87	7.72	7.56	7.40	7.23	7.09	6.88
7	5.59	4.74	4.35	4.12	3.97	3.87	3.79	3.73	3.68	3.64	3.57	3.51	3.44	3.38	3.32	3.23
	12.2	9.55	8.45	7.85	7.46	7.19	6.99	6.84	6.72	6.62	6.47	6.31	6.16	5.99	5.86	5.65
8	5.32	4.46	4.07	3.84	3.69	3.58	3.50	3.44	3.39	3.35	3.28	3.22	3.15	3.08	3.02	2.93
	11.3	8.65	7.59	7.01	6.63	6.37	6.18	6.03	5.91	5.81	5.67	5.52	5.36	5.20	5.07	4.86
9	5.12	4.26	3.86	3.63	3.48	3.37	3.29	3.23	3.18	3.14	3.07	3.01	2.94	2.86	2.80	2.71
	10.6	8.02	6.99	6.42	6.06	5.80	5.61	5.47	5.35	5.26	5.11	4.96	4.81	4.65	4.52	4.31
10	4.96	4.10	3.71	3.48	3.33	3.22	3.14	3.07	3.02	2.98	2.91	2.85	2.77	2.70	2.64	2.54
	10.0	7.56	6.55	5.99	5.64	5.39	5.20	5.06	4.94	4.85	4.71	4.56	4.41	4.25	4.12	3.91
11	4.84	3.98	3.59	3.36	3.20	3.09	3.01	2.95	2.90	2.85	2.79	2.72	2.65	2.57	2.51	2.40
	9.65	7.21	6.22	5.67	5.32	5.07	4.89	4.74	4.63	4.54	4.40	4.25	4.10	3.94	3.81	3.60
12	4.75	3.89	3.49	3.26	3.11	3.00	2.91	2.85	2.80	2.75	2.69	2.62	2.54	2.47	2.40	2.30
	9.33	6.93	5.95	5.41	5.06	4.82	4.64	4.50	4.39	4.30	4.16	4.01	3.86	3.70	3.57	3.36
13	4.67	3.81	3.41	3.18	3.03	2.92	2.83	2.77	2.71	2.67	2.60	2.53	2.46	2.38	2.31	2.21
	9.07	6.70	5.74	5.21	4.86	4.62	4.44	4.30	4.19	4.10	3.96	3.82	3.66	3.51	3.38	3.17
14	4.60	3.74	3.34	3.11	2.96	2.85	2.76	2.70	2.65	2.60	2.53	2.46	2.39	2.31	2.24	2.13
	8.86	6.51	5.56	5.04	4.69	4.46	4.28	4.14	4.03	3.94	3.80	3.66	3.51	3.35	3.22	3.00
15	4.54	3.68	3.29	3.06	2.90	2.79	2.71	2.64	2.59	2.54	2.48	2.40	2.33	2.25	2.18	2.07
	8.68	6.36	5.42	4.89	4.56	4.32	4.14	4.00	3.89	3.80	3.67	3.52	3.37	3.21	3.08	2.87
16	4.49	3.63	3.24	3.01	2.85	2.74	2.66	2.59	2.54	2.49	2.42	2.35	2.28	2.19	2.12	2.01
	8.53	6.23	5.29	4.77	4.44	4.20	4.03	3.89	3.78	3.69	3.55	3.41	3.26	3.10	2.97	2.75
17	4.45	3.59	3.20	2.96	2.81	2.70	2.61	2.55	2.49	2.45	2.38	2.31	2.23	2.15	2.08	1.96
	8.40	6.11	5.18	4.67	4.34	4.10	3.93	3.79	3.68	3.59	3.46	3.31	3.16	3.00	2.87	2.65
18	4.41	3.55	3.16	2.93	2.77	2.66	2.58	2.51	2.46	2.41	2.34	2.27	2.19	2.11	2.04	1.92
	8.29	6.01	5.09	4.58	4.25	4.01	3.84	3.71	3.60	3.51	3.37	3.23	3.08	2.92	2.78	2.57
19	4.38	3.52	3.13	2.90	2.74	2.63	2.54	2.48	2.42	2.38	2.31	2.23	2.16	2.07	2.00	1.88
	8.18	5.93	5.01	4.50	4.17	3.94	3.77	3.63	3.52	3.43	3.30	3.15	3.00	2.84	2.71	2.49
20	4.35	3.49	3.10	2.87	2.71	2.60	2.51	2.45	2.39	2.35	2.28	2.20	2.12	2.04	1.97	1.84
	8.10	5.85	4.94	4.43	4.10	3.87	3.70	3.56	3.46	3.37	3.23	3.09	2.94	2.78	2.64	2.42
25	4.24	3.39	2.99	2.76	2.60	2.49	2.40	2.34	2.28	2.24	2.16	2.09	2.01	1.92	1.84	1.71
	7.77	5.57	4.68	4.18	3.85	3.63	3.46	3.32	3.22	3.13	2.99	2.85	2.70	2.54	2.40	2.17
30	4.17	3.32	2.92	2.69	2.53	2.42	2.33	2.27	2.21	2.16	2.09	2.01	1.93	1.84	1.76	1.62
	7.56	5.39	4.51	4.02	3.70	3.47	3.30	3.17	3.07	2.98	2.84	2.70	2.55	2.39	2.25	2.01
35	4.12	3.27	2.87	2.64	2.49	2.37	2.29	2.22	2.16	2.11	2.04	1.96	1.88	1.79	1.70	1.56
	7.42	5.27	4.40	3.91	3.59	3.37	3.20	3.07	2.96	2.88	2.74	2.60	2.44	2.28	2.14	1.89
40	4.08	3.23	2.84	2.61	2.45	2.34	2.25	2.18	2.12	2.08	2.00	1.92	1.84	1.74	1.66	1.51
	7.31	5.18	4.31	3.83	3.51	3.29	3.12	2.99	2.89	2.80	2.66	2.52	2.37	2.20	2.06	1.80

Df in denominator	Df in numerator															
	1	2	3	4	5	6	7	8	9	10	12	15	20	30	50	∞
45	4.06	3.20	2.81	2.58	2.42	2.31	2.22	2.15	2.10	2.05	1.97	1.89	1.81	1.71	1.63	1.47
	7.23	5.11	4.25	3.77	3.45	3.23	3.07	2.94	2.83	2.74	2.61	2.46	2.31	2.14	2.00	1.74
50	4.03	3.18	2.79	2.56	2.40	2.29	2.20	2.13	2.07	2.03	1.95	1.87	1.78	1.69	1.60	1.44
	7.17	5.06	4.20	3.72	3.41	3.19	3.02	2.89	2.78	2.70	2.56	2.42	2.27	2.10	1.95	1.68
60	4.00	3.15	2.76	2.53	2.37	2.25	2.17	2.10	2.04	1.99	1.92	1.84	1.75	1.65	1.56	1.39
	7.08	4.98	4.13	3.65	3.34	3.12	2.95	2.82	2.72	2.63	2.50	2.35	2.20	2.03	1.88	1.60
80	3.96	3.11	2.72	2.49	2.33	2.21	2.13	2.06	2.00	1.95	1.88	1.79	1.70	1.60	1.51	1.32
	6.96	4.88	4.04	3.56	3.26	3.04	2.87	2.74	2.64	2.55	2.42	2.27	2.12	1.94	1.79	1.49
100	3.94	3.09	2.70	2.46	2.31	2.19	2.10	2.03	1.97	1.93	1.85	1.77	1.68	1.57	1.48	1.28
	6.90	4.82	3.98	3.51	3.21	2.99	2.82	2.69	2.59	2.50	2.37	2.22	2.07	1.89	1.74	1.43
120	3.92	3.07	2.68	2.45	2.29	2.18	2.09	2.02	1.96	1.91	1.83	1.75	1.66	1.55	1.46	1.25
	6.85	4.79	3.95	3.48	3.17	2.96	2.79	2.66	2.56	2.47	2.34	2.19	2.03	1.86	1.70	1.38
∞	3.84	3.00	2.60	2.37	2.21	2.10	2.01	1.94	1.88	1.83	1.75	1.67	1.57	1.46	1.35	1.00
	6.63	4.61	3.78	3.32	3.02	2.80	2.64	2.51	2.41	2.32	2.18	2.04	1.88	1.70	1.52	1.00

APPENDIX 7   THE CHI-SQUARE DISTRIBUTION

The table gives the critical values of χ^2 in a two-tailed/non-directional test, for different numbers of degrees of freedom (df). For significance, the calculated value must be *greater than or equal to* the critical value.

df	0.20	0.10	0.05	0.025	0.01	0.001
			Significance level			
1	1.64	2.71	3.84	5.02	6.64	10.83
2	3.22	4.61	5.99	7.38	9.21	13.82
3	4.64	6.25	7.82	9.35	11.34	16.27
4	5.99	7.78	9.49	11.14	13.28	18.47
5	7.29	9.24	11.07	12.83	15.09	20.52
6	8.56	10.64	12.59	14.45	16.81	22.46
7	9.80	12.02	14.07	16.01	18.48	24.32
8	11.03	13.36	15.51	17.53	20.09	26.12
9	12.24	14.68	16.92	19.02	21.67	27.88
10	13.44	15.99	18.31	20.48	23.21	29.59
11	14.63	17.28	19.68	21.92	24.72	31.26
12	15.81	18.55	21.03	23.34	26.22	32.91
13	16.98	19.81	22.36	24.74	27.69	34.53
14	18.15	21.06	23.68	26.12	29.14	36.12
15	19.31	22.31	25.00	27.49	30.58	37.70
16	20.47	23.54	26.30	28.85	32.00	39.25
17	21.61	24.77	27.59	30.19	33.41	40.79
18	22.76	25.99	28.87	31.53	34.81	42.31
19	23.90	27.20	30.14	32.85	36.19	43.82
20	25.04	28.41	31.41	34.17	37.57	45.31
21	26.17	29.62	32.67	35.48	38.93	46.80
22	27.30	30.81	33.92	36.78	40.29	48.27
23	28.43	32.01	35.17	38.08	41.64	49.73
24	29.55	33.20	36.42	39.36	42.98	51.18
25	30.68	34.38	37.65	40.65	44.31	52.62
26	31.79	35.56	38.89	41.92	45.64	54.05
27	32.91	36.74	40.11	43.19	46.96	55.48
28	34.03	37.92	41.34	44.46	48.28	56.89
29	35.14	39.09	42.56	45.72	49.59	58.30
30	36.25	40.26	43.77	46.98	50.89	59.70
40	47.27	51.81	55.76	59.34	63.69	73.40
50	58.16	63.17	67.50	71.42	76.15	86.66
60	68.97	74.40	79.08	83.30	88.38	99.61
70	79.71	85.53	90.53	95.02	100.4	112.3

APPENDIX 8   THE PEARSON PRODUCT–MOMENT CORRELATION COEFFICIENT

The table gives the critical values of the Pearson product–moment correlation coefficient, r, for different numbers of pairs of observations, N. For significance, the calculated value of r must be *greater than or equal to* the critical value.

	Significance level: two-tailed / non-directional			
	0.20	0.10	0.05	0.01
	Significance level: one-tailed / directional			
N	0.10	0.05	0.025	0.005
3	0.951	0.988	0.997	1.000
4	0.800	0.900	0.950	0.990
5	0.687	0.805	0.878	0.959
6	0.608	0.729	0.811	0.917
7	0.551	0.669	0.754	0.875
8	0.507	0.621	0.707	0.834
9	0.472	0.582	0.666	0.798
10	0.443	0.549	0.632	0.765
11	0.419	0.521	0.602	0.735
12	0.398	0.497	0.576	0.708
13	0.380	0.476	0.553	0.684
14	0.365	0.458	0.532	0.661
15	0.351	0.441	0.514	0.641
16	0.338	0.426	0.497	0.623
17	0.327	0.412	0.482	0.606
18	0.317	0.400	0.468	0.590
19	0.308	0.389	0.456	0.575
20	0.299	0.378	0.444	0.561
21	0.291	0.369	0.433	0.549
22	0.284	0.360	0.423	0.537
23	0.277	0.352	0.413	0.526
24	0.271	0.344	0.404	0.515
25	0.265	0.337	0.396	0.505
26	0.260	0.330	0.388	0.496
27	0.255	0.323	0.381	0.487
28	0.250	0.317	0.374	0.479
29	0.245	0.311	0.367	0.471
30	0.241	0.306	0.361	0.463
40	0.207	0.264	0.312	0.403
50	0.184	0.235	0.279	0.361
60	0.168	0.214	0.254	0.330
70	0.155	0.198	0.235	0.306
80	0.145	0.185	0.220	0.286
90	0.136	0.174	0.207	0.270
100	0.129	0.165	0.197	0.256
200	0.091	0.117	0.139	0.182

APPENDIX 9   THE SPEARMAN RANK CORRELATION COEFFICIENT

The table gives the critical values of the Spearman rank correlation coefficient, ρ, for different numbers of pairs of observations, N.

	Significance level: two-tailed/non-directional			
	0.20	*0.10*	*0.05*	*0.01*
	Significance level: one-tailed/directional			
N	*0.10*	*0.05*	*0.025*	*0.005*
5	0.800	0.900	1.000	–
6	0.657	0.829	0.886	1.000
7	0.571	0.714	0.786	0.929
8	0.524	0.643	0.738	0.881
9	0.483	0.600	0.700	0.833
10	0.455	0.564	0.648	0.794
11	0.427	0.536	0.618	0.755
12	0.406	0.503	0.587	0.727
13	0.385	0.484	0.560	0.703
14	0.367	0.464	0.538	0.679
15	0.354	0.446	0.521	0.654
16	0.341	0.429	0.503	0.635
17	0.328	0.414	0.488	0.618
18	0.317	0.401	0.472	0.600
19	0.309	0.391	0.460	0.584
20	0.299	0.380	0.447	0.570
21	0.292	0.370	0.436	0.556
22	0.284	0.361	0.425	0.544
23	0.278	0.353	0.416	0.532
24	0.271	0.344	0.407	0.521
25	0.265	0.337	0.398	0.511
26	0.259	0.331	0.390	0.501
27	0.255	0.324	0.383	0.492
28	0.250	0.318	0.375	0.483
29	0.245	0.312	0.368	0.475
30	0.240	0.306	0.362	0.467
35	0.222	0.283	0.335	0.433
40	0.207	0.264	0.313	0.405
45	0.194	0.248	0.294	0.382
50	0.184	0.235	0.279	0.363
55	0.175	0.224	0.266	0.346
60	0.168	0.214	0.255	0.331

Answers to exercises

CHAPTER 1
1. (a) yes (b) no (c) yes (d) no (e) yes.
2. (a) ANOVA, (b) matched pairs t test, (c) non-parametric test, such as median or rank sums, (d) non-parametric test (since mean is better measure of central tendency than the mean), such as median or rank sums.
3. chi-square = 100, df = 4, significance < 0.001.

CHAPTER 2
1. 13/21 = 0.62.
2. (a) 1.759 bits (b) 2.322 bits (c) 0.758 (d) 0.242.
3. The = article, *stock* = adjective, *market* = noun, *run* = verb, *continues* = verb.

CHAPTER 3
1. Sundanese and Malay merge at 0.82.
   Javanese and Madurese merge at 0.75.
   Sundanese-Malay and Javanese-Madurese merge at 0.4325.
2. (a) 3 (b) 0.4 (c) 0.5.
3. (a) 3 (b) 5. $D_3 = 64$.
4. (a) *solut-* (b) *solubl-* (c) *solut-* (d) *solut-*.
   solution, solve and *solvable* form an equivalence class.

CHAPTER 4
1. *machine.*
2. (a) *cat* (b) *cat* and *white* are equally likely.
3. *a number of.*

CHAPTER 5
1. York, with odds 1.086 to 1.
2. 2,222
3. 10
4. 0.447

Bibliography

Adamson, G. W. and Boreham, J. (1974), 'The Use of an Association Measure Based on Character Structure to Identify Semantically Related Pairs of Words and Document Titles', *Information Storage and Retrieval*, 10, pp. 253–60.

Agresti, A. (1990), 'Categorical Data Analysis', Bognor Regis: Wiley.

Ahmad, F., Yusoff, M. and Sembok, T. M. T. (1996), 'Experiments with a Stemming Algorithm for Malay Words', *Journal of the American Society of Information Science (JASIS)*, 47 (12), pp. 909–18.

Alavi, S. M. (1994), 'Traditional and Modern Approaches to Test Data Analysis', in Alavi, S. M. and de Mejia, A.-M. (eds), *Data Analysis in Applied Linguistics*, Department of Linguistics, Lancaster University.

Allen, J. R. (1974), 'On the Authenticity of the Baligant Episode in the *Chanson de Roland*', in Mitchell, J. L. (ed.), *Computers in the Humanities*, Edinburgh: Edinburgh University Press.

Alt, M. (1990), *Exploring Hyperspace*, Maidenhead: McGraw-Hill.

Altenberg, B. and Eeg-Olofsson, M. (1990), 'Phraseology in Spoken English: Presentation of a Project', in Aarts, J. and Meijs, W. (eds), *Theory and Practice in Corpus Linguistics*, Amsterdam and Atlanta: Rodopi.

Anderberg, M. R. (1973), *Cluster Analysis for Applications*, New York: Academic Press.

Angell, R. C., Freund, G. E. and Willett, P. (1983), 'Automatic Spelling Correction Using a Trigram Similarity Measure', *Information Processing and Management*, 19(4), pp. 255–61.

Ashton, R., (1996), 'Revenge of the Trainspotters', *Focus*, September, pp. 32–5.

Atwell, E. (1987), 'Constituent-Likelihood Grammar', in Garside, R., Leech, G. and Sampson, G. (eds), *The Computational Analysis of English: A Corpus-Based Approach*, London and New York: Longman.

Atwell, E. and Elliott, S. (1987), 'Dealing with Ill-Formed English Text', in Garside, R., Leech, G. and Sampson, G. (eds), *The Computational Analysis of English: A Corpus-Based Approach*, London and New York: Longman.

Austin, W. B. (1966), 'The Posthumous Greene Pamphlets: A Computerised Study', *Shakespeare Newsletter*, 16, p. 45.

Baayen, H. (1992), 'Statistical Models for Word Frequency Distributions: A Linguistic

Evaluation', *Computers and the Humanities*, 26, pp. 347–63.

Baayen, H., van Halteren, H. and Tweedie, F. (1996), 'Outside the Cave of Shadows: Using Syntactic Annotation to Enhance Authorship Attribution', *Literary and Linguistic Computing*, 11(3).

Bailey R.W. (1969), 'Statistics and Style: A Historical Survey', in Dolezel, L. and Bailey, R.W. (eds), *Statistics and Style*, New York: American Elsevier Publishing Company.

Baillie, W. M. (1974), 'Authorship Attribution in Jacobean Dramatic Texts', in Mitchell, J. L. (ed.), *Computers in the Humanities*, Edinburgh: Edinburgh University Press.

Bar-Hillel, Y. (1964), *Language and Information*, Reading, Massachusetts: Addison Wesley Publishing Company.

Barnbrook, G. (1996), *Language and Computers: A Practical Introduction to the Computer Analysis of Language*, Edinburgh: Edinburgh University Press.

Baum, L. E, and Eagon, J. A. (1967), 'An Inequality with Applications to Statistical Estimation for Probabilistic Functions to Markov Processes and to a Model for Ecology', *American Mathematical Society Bulletin*, 73, pp. 360–3.

Baum, L. E., Petrie, T., Soules, G. and Weiss, N. (1970), 'A Maximization Technique Occurring in the Statistical Analysis of Probabilistic Functions of Markov Chains', *The Annals of Mathematical Statistics*, 41(1), pp. 164–71.

Berry-Rogghe, G. L. M. (1973), 'The Computation of Collocations and their Relevance in Lexical Studies', in Aitken, A. J., Bailey, R. and Hamilton-Smith, N. (eds), *The Computer and Literary Studies*, Edinburgh: Edinburgh University Press.

Berry-Rogghe, G. L. M., (1974), 'Automatic Identification of Phrasal Verbs', in Mitchell, J. L. (ed.), *Computers in the Humanities*, Edinburgh: Edinburgh University Press.

Biber, D. (1988), *Variation Across Speech and Writing*, Cambridge: Cambridge University Press.

Biber, D. (1993), 'Co-occurrence Patterns Among Collocations: A Tool for Corpus-Based Lexical Knowledge Acquisition', *Computational Linguistics* 19(3), pp. 531–8.

Biber, D. (1995), *Dimensions of Register Variation*, Cambridge: Cambridge University Press.

Biber, D. and Finegan, E. (1986), 'An Initial Typology of English Text Types', in Aarts, J. and Meijs, W. (eds), *Corpus Linguistics II*, Amsterdam: Rodopi.

Brainerd, B. (1974), *Weighing Evidence in Language and Literature: A Statistical Approach*, Toronto and Buffalo: University of Toronto Press.

Bross, I. D. J., Shapiro, P. A. and Anderson, B. B. (1972), 'How Information is Carried in Scientific Sublanguages', *Science*, 176, pp. 1303–7.

Brown, P. F., Cocke, J., Della Pietra, S. A., Della Pietra, V. J., Jelinek, F., Mercer R. L. and Roosin, P. S. (1988), 'A Statistical Approach to Language Translation', in *Proceedings of the 12th International Conference on Computational Linguistics (Coling 88)*, Budapest.

Brown, P. F., Cocke, J., Della Pietra, S. A., Della Pietra, V. J., Jelinek, F., Lafferty, J. D., Mercer, R. L. and Roosin, P. S. (1990), 'A Statistical Approach to Machine Translation', *Computational Linguistics*, 16(2).

Brown, P., Lai, J. and Mercer, R. (1991), 'Aligning Sentences in Parallel Corpora', in *Proceedings of the 29th Annual Meeting of the ACL*, pp. 169–76.

Brown University Standard Corpus of Present Day American English (see Kucera and Francis).

Butler, C. (1985a), *Statistics in Linguistics*, Oxford: Basil Blackwell.

Butler, C. (1985b), *Computers in Linguistics*, Oxford: Basil Blackwell.

Campbell (1867) (see Kenny 1982).

Canter, D. (1992), 'An Evaluation of the "Cusum" Stylistic Analysis of Confessions', *Expert Evidence*, 1(3), pp. 93–9.

Carroll, J. B. (1970), 'An Alternative to Juilland's Usage Coefficient for Lexical Frequencies and a Proposal for a Standard Frequency Index (SFI)', *Computer Studies in the Humanities and Verbal Behaviour*, 3(2), pp. 61–5.

Chadwick, J. (1958), *The Decipherment of Linear B*, Cambridge: Cambridge University Press.

Church, K., Gale, W., Hanks, P. and Hindle, D. (1991), 'Using Statistics in Lexical Analysis', in Zernik, U. (ed.), *Lexical Acquisition: Exploiting On-line Resources to Build a Lexicon*, Hillsdale, NJ: Lawrence Erlbaum Associates.

Church, K. W. and Hanks, P. (1990), 'Word Association Norms, Mutual Information and Lexicography', *Computational Linguistics*, 16(1), pp. 22–9.

Clapham, C. (1996), *The Concise Oxford Dictionary of Mathematics*, Oxford and New York: Oxford University Press.

Coggins, J. M., (1983), 'Dissimilarity Measures for Clustering Strings', in Sankoff, D. and Kruskal, J. B. (eds), *Time Warps, String Edits, and Macromolecules: The Theory and Practice of Sequence Comparison*, Reading, MA: Addison-Wesley Publishing Company, pp. 311–22.

Cutting, D., Kupiek, J., Pedersen J. and Sibun, P. (1992), *A Practical Part-of-Speech Tagger*, Palo Alto, CA: Xerox Palo Alto Research Center.

Dagan, I., Itai, A. and Schwall, U. (1991), 'Two Languages are more Informative than One', in *Proceedings of the 29th Annual Meeting of the Association for Computational Linguistics*, pp. 130–7.

Daille, B., (1994), 'Extraction Automatique de Noms Composés Terminologiques', Ph.D. thesis, University of Paris 7.

Daille, B. (1995), 'Combined Approach for Terminology Extraction: Lexical Statistics and Linguistic Filtering', UCREL *Technical Papers*, Volume 5, Department of Linguistics, University of Lancaster, 1995.

Damerau, F. J., (1964), 'A Technique for the Computer Detection and Correction of Spelling Errors', *Communications of the ACM*, 7, pp. 171–6.

Damerau, F. J. (1971), *Markov Models and Linguistic Theory*, The Hague: Mouton.

Damerau, F. J. (1975), 'The Use of Function Word Frequencies as Indicators of Style', *Computing in the Humanities*, 9, pp. 271–80.

de Haan, P. and van Hout, R. (1986), 'A Loglinear Analysis of Syntactic Constraints on Postmodifying Clauses', in Aarts, J. and Meijs, W. (eds), *Corpus Linguistics II: New Studies in the Analysis and Exploration of Computer Corpora*, Amsterdam: Rodopi.

de Tollenaere, F. (1973), 'The Problem of the Context in Computer-Aided Lexicography' in Aitken, A. J., Bailey, R. W. and Hamilton-Smith, N. (eds), *The Computer and Literary Studies*, Edinburgh: Edinburgh University Press.

Dempster, A. P., Laird, N. M. and Rubin, D. B. (1997), 'Maximum Likelihood from Incomplete Data via the EM Algorithm', *Journal of the Royal Statistical Society*, Series B (Methodological), 39, pp. 1–22.

Denning, D. E. R., (1982), *Cryptography and Data Security*, Reading, MA: Addison Wesley.

Derwent Drug File Thesaurus, in *Ringdoc/Vetdoc Instruction Bulletin*, numbers UDBIA and UDB2 PT 1–3, *Thesaurus and Thesaurus Appendices* (1986), (Fourth Edition), London: Derwent Publications Ltd.

Dewey, G. (1923), *Relative Frequency of English Speech Sounds*, Cambridge, MA: Harvard University Press.

Dice, L. R. (1945), 'Measures of the Amount of Ecologic Association Between Species', *Geology*, 26, pp. 297–302.

Dunning, T. (1993), 'Accurate Methods for the Statistics of Surprise and Coincidence', *Computational Linguistics*, 19(1), pp. 61–74.

Dyen, I. (1962), 'The Lexicostatistical Classification of the Malayo-Polynesian Languages, *Language*, 38, pp. 38–46.

Dyen, I., James, A. T. and Cole, J. W. L. (1967), 'Language Divergence and Estimated Word Retention Rate', *Language*, 43(1), pp. 150–71.

Eastment, H. T. and Krzanowski, W. J., (1982), 'Cross-Validatory Choice of the Number of Components from a Principal Components Analysis', *Technometrics*, 24, pp. 73–7.

Edmundson, H. P. (1963), 'A Statistician's View of Linguistic Models and Language Data Processing', in Garvin, P. L. (ed.), *Natural Language and the Computer*, New York: McGraw Hill.

Ellegård, A. (1959), 'Statistical Measurement of Linguistic Relationship', *Language*, 35(2), pp. 131–56.

Ellegård, A. (1962), *A Statistical Method for Determining Authorship*, Gothenburg: Acta Universitas Gothoburgensis.

Ellison, J. W. (1965), 'Computers and the Testaments', in *Computers for the Humanities?*, New Haven, CT: Yale University Press, pp. 72–4 (see Oakman 1980, p. 140).

Ertet (no date given) (see Kenny 1982).

Évrard, Étienne (1966), 'Étude statistique sur les affinités de cinquante-huit dialectes bantous', in *Statistique et analyse linguistique*, Centres d'Études Supérieures Spécialisés, Université de Strasbourg, Paris: Presses Universitaires de France, pp. 85–94.

Forney, G. D. Jr., (1973), 'The Viterbi Algorithm', *Proceedings of the IEEE*, 61(3), p. 268.

Francis, I. S. (1966), 'An Exposition of a Statistical Approach to the Federalist Dispute', in Leed, J. (ed.), *The Computer and Literary Style*, Kent, OH: Kent State University Press.

Gale, W. A. and Church, K. W. (1991), 'Concordances for Parallel Texts', in *Proceedings of the Seventh Annual Conference of the UW Centre for the New OED and Text Research Using Corpora*, Oxford, pp. 40–62.

Gale, W. A. and Church, K. W. (1993), 'A Program for Aligning Sentences in Bilingual Corpora', *Computational Linguistics*, 19(1), pp. 75–102.

Gale, W. A., Church K. W. and Yarowsky, D. (1992), 'A Method for Disambiguating Word Senses in a Large Corpus', *Computers and the Humanities*, 26 (5–6), pp. 415–39.

Garside, R. (1987), 'The CLAWS Word Tagging System', in Garside, R., Leech, G. and Sampson, G. (eds), *The Computational Analysis of English*, London and New York: Longman.

Garside, R., Leech, G. and McEnery, A. (eds) (1997), *Corpus Annotation*, London: Addison-Wesley-Longman.

Gaussier, E. and Langé, J.-M. (1994), 'Some Methods for the Extraction of Bilingual Terminology', in Jones, D. (ed.) *Proceedings of the International Conference on New Methods in Language Processing* (NeMLaP), 14–16 September 1994, UMIST, Manchester, pp. 242–7.

Gaussier, E. and Langé, J.-M. and Meunier, F. (1992), 'Towards Bilingual Terminology', 19th International Conference of the Association for Literary and Linguistic Computing, 6–9 April 1992, Christ Church Oxford, in *Proceedings of the Joint ALLC/ACH Conference*, Oxford: Oxford University Press, pp. 121–4.

Geake, E. (1994), 'Aethelburg Knew Me', *New Scientist*, 9 April.

Geffroy, A., Guilhaumou, J., Hartley, A. and Salem, A. (1976), 'Factor Analysis and Lexicometrics: Shifters in Some Texts of the French Revolution (1793–1794)', in Jones, A. and Churchhouse, R. F. (eds), *The Computer in Literary and Linguistic Studies*, Cardiff: University of Wales Press.

Geffroy, A., Lafon, P., Seidel, G. and Tournier, M. (1973), 'Lexicometric Analysis of Co-occurrences', in Aitken, A. J., Bailey, R. W. and Hamilton-Smith, N. (eds), *The Computer and Literary Studies*, Edinburgh: Edinburgh University Press.

Gilbert, N. (1993), *Analysing Tabular Data*, London: UCL Press.

Gledhill, C. (1996), 'Science as a Collocation: Phraseology in Cancer Research Articles', in Botley, S., Glass, J., McEnery, A. and Wilson, W. (eds), *Proceedings of Teaching and Language Corpora 1996*, UCREL Technical Papers, Volume 9 (Special Issue), Department of Linguistics, Lancaster University, pp. 108–26.

Goldman-Eisler, F. G. (1958), 'Speech Production and the Predictability of Words in Context', *Quarterly Journal of Experimental Psychology*, 10, 96–109.

Hall, P. A. and Dowling, G. R. (1980), 'Approximate String Matching', ACM *Computing Surveys*, 12(4), p. 381.

Halliday, M. A. K. (1991), 'Corpus Studies and Probabilistic Grammar', in Aijmer, K. and Altenberg, B. (eds), *English Corpus Linguistics*, London and New York: Longman.

Hanlon, S. and Boyle,. R. D. (1992), 'Syntactic Knowledge in Word Level Text Recognition', in Beale, R. and Findlay, J. (eds), *Neural Networks and Pattern Recognition in HCI*, Hemel Hempstead: Ellis Horwood.

Hanks, P. (ed.), (1986), *The Collins English Dictionary*, 2nd edn, Glasgow: William Collins and Sons.

Hann, M. N. (1973), 'The Statistical Force of Random Distribution', ITL, 20, pp. 31–44.

Haskel, P. I. (1971), 'Collocations as a Measure of Stylistic Variety', in Wisbey, R. A. (ed.), *The Computer in Literary and Linguistic Research*, Cambridge: Cambridge University Press.

Hatch, Evelyn M. and Lazaraton, Anne (1991), *The Research Manual: Design and Statistics for Applied Linguistics*, Boston, MA: Heinle and Heinle.

Hays, D. G., (1967), *Introduction to Computational Linguistics*, London: MacDonald and Co.

Herdan, G. (1962), *The Calculus of Linguistic Observations*, The Hague: Mouton and Co.

Hickey, R. (1993), 'Corpus Data Processing with Lexa', ICAME Journal 17, pp. 73–95.

Hirschberg, D. S. (1983), 'Recent Results on the Complexity of Common Subsequence Problems', in Sankoff, D. and Kruskal, J. B. (eds), *Time Warps, String Edits, and Macromolecules: The Theory and Practice of Sequence Comparison*, Reading, MA: Addison-Wesley.

Hockey, S. (1980), *A Guide to Computer Applications in the Humanities*, London: Duckworth.

Hofland, K., (1991), 'Concordance Programs for Personal Computers', in Johansson, S. and Stenström, A.-B. (eds), *English Computer Corpora*, Berlin and New York: Mouton de Gruyter.

Holmes, D. I. (1992), 'A stylometric analysis of Mormon scripture and related texts', *Journal of the Royal Statistical Society*, A155(1), pp. 91–120.

Holmes, D. I. (1994), 'Authorship Attribution', *Computers and the Humanities*, 28, pp. 87–106.

Holmes, D. I. and Forsyth, R. S. (1995), 'The Federalist Revisited: New Directions in Authorship Attribution', *Literary and Linguistic Computing*, 10, pp. 111–27.

Holmes, D. I. and Tweedie, F. J. (1995), 'Forensic Stylometry: A Review of the Cusum Controversy', *Revue Informatique et Statistique dans les Sciences Humaines*, pp. 19–47.

Hood-Roberts, A. H. (1965), *A Statistical Linguistic Analysis of American English*, The Hague: Mouton and Co.

Horvath, B. M. (1985), *Variation in Australian English*, Cambridge: Cambridge University Press.

Hosaka, J., Seligman, M. and Singer, H. (1994), 'Pause as a Phrase Demarcator for Speech and Language Processing', *Proceedings of COLING 94*, vol. 2, pp. 987–91.

Iida, H. (1995), 'Spoken Dialogue Translation Technologies and Speech Translation', The 5th Machine Translation Summit, Luxembourg, 10–13 July.

Jelinek, F. (1985), 'Self-Organized Language Modeling for Speech Recognition', IBM Research Report, Continuous Speech Recognition Group, IBM T. J. Watson Research Center, Yorktown Heights, New York.

Jelinek, F. (1990), 'Self-Organized Language Modeling for Speech Recognition', in Waibel, A. and Lee, K.F. (eds), *Readings in Speech Recognition*, New York: Morgan Kaufman Publishers.

Jolicoeur, P. and Mossiman, J. E. (1960), 'Size and Shape Variation in the Painted Turtle: a Principal Component Analysis', *Growth*, 24, pp. 339–54.

Joseph, D. M. and Wong, R. L. (1979), 'Correction of Misspellings and Typographical Errors in a Free-Text Medical English Information Storage and Retrieval System', *Methods of Information in Medicine*, 18(4), pp. 228–34.

Juilland, A., Brodin, D. and Davidovitch, C. (1970), *Frequency Dictionary of French Words*, The Hague: Mouton.

Juola, P., Hall, C. and Boggs, A. (1994), 'Corpus Based Morphological Segmentation by Entropy Changes', in Moneghan, A. (ed.), *CS-NLP Proceedings*, Dublin City University, 6–8 July.

Kahn, D. (1966), *The Codebreakers*, London: Weidenfeld and Nicolson.

Kaufman, L. and Rousseeuw, P. J. (1990), *Finding Groups in Data: An Introduction to Data Analysis*, New York: John Wiley & Sons.

Kay, M. and Röscheisen, M. (1993), 'Text–Translation Alignment', *Computational Linguistics*, 19(1), pp. 121–42.

Kenny, A. J. P. (1977), 'The Stylometric Study of Aristotle's Ethics', in Luisignan, S. and North, J. S. (eds), *Computing and the Humanities*, Waterloo: University of Waterloo Press.

Kenny, A. J. P. (1978), *The Aristotlean Ethics*, Oxford: Oxford University Press.

Kenny, A. J. P. (1982), *The Computation of Style*, Oxford: Pergamon Press.

Kessler, B. (1995), 'Computational Dialectology in Irish Gaelic', Seventh Conference of the European Chapter of the Association for Computational Linguistics, University College Dublin, 27–31 March.

Kibaroglu, M. O. and Kuru, S. (1991), 'A Left to Right Morphological Parser for Turkish', in Baray, M. and Ozguc, B. (eds), *Computer and Information Science*, Amsterdam: Elsevier.

Kilgarriff, A. (1996a), 'Which Words are Particularly Characteristic of Text? A Survey of Statistical Approaches', Information Technology Research Institute, University of Brighton, 6 March.

Kilgarriff, A. (1996b), 'Using Word Frequency Lists to Measure Corpus Homogeneity and Similarity between Corpora', Information Technology Research Institute, University of Brighton, 18 April.

Kilgarriff, A. (1996c), 'Corpus Similarity and Homogeneity via Word Frequency', *EURALEX Proceedings*, Gothenburg, Sweden, August.

Kirk, J. M. (1994), 'Taking a Byte at Corpus Linguistics', in Flowerdew, L. and Tong, A. K. K. (eds), 'Entering Text', Language Centre, The Hong Kong University of Science and Technology.

Kita, K., Kato, Y., Omoto, T. and Yano, Y. (1994), 'Automatically Extracting Collocations from Corpora for Language Learning', in Wilson, A. and McEnery, A. (eds), *UCREL Technical Papers*, Volume 4 (Special Issue), *Corpora in Language Education and Research, A Selection of Papers from Talc94*, Department of Linguistics, Lancaster University.

Kjellmer, G. (1984), 'Some Thoughts on Collocational Distinctiveness', in Aarts, J. amd Meijs, W. (eds), *Corpus Linguistics*, Amsterdam: Rodopi.

Kjellmer, G. (1990), 'Patterns of Collocability', in Aarts, J. and Meijs, W. (eds), *Theory and Practice in Corpus Linguistics*, Amsterdam and Atlanta: Rodopi.

Kjetsaa, G. (1976), 'Storms on the Quiet Don: A Pilot Study', *Scando-Slavica*, 22, pp. 5–24.

Knuth, D. E. (1973), *The Art of Computer Programming: Volume 3: Sorting and Searching*, Reading, MA: Addison Wesley.

Kober, A. (1945), 'Evidence of Inflection in the "Chariot" Tablets from Knossos', *American Journal of Archaeology*, XLVIII, January–March 1944, pp. 64–75.

Köster, P., (1971), 'Computer Stylistics: Swift and Some Contemporaries', in Wisbey, R. A. (ed.), *The Computer in Literary and Linguistic Research*, Cambridge: Cambridge University Press.

Kondratov, A. M. (1969), 'Information Theory and Poetics: the Entropy of Russian Speech Rhythm', in Dolezel, L. and Bailey, R. W. (eds), *Statistics and Style*, New York: Elsevier.

Krause, P. and Clark, D. (1993), *Representing Uncertain Knowledge*, Dordrecht: Kluwer Academic Publishers.

Kruskal, J. B. (1983), 'An Overview of Sequence Comparison', in Sankoff, D. and Kruskal, J. B. (eds), *Time Warps, String Edits, and Macromolecules: The Theory and Practice of Sequence Comparison*, Reading, MA: Addison-Wesley.

Kucera, H. and Francis, W. N. (1967), *Computational Analysis of Present Day American English*, Providence, RI: Brown University Press.

Lafon, P. (1984), *Dépouillements et statistiques en lexicométrie*, Geneva and Paris: Slatkine-Champion.

Lancashire, I. (1995), 'Computer Tools for Cognitive Stylistics', in Nissan, E. and Schmidt, K. M. (eds), *From Information to Knowledge*, Oxford: Intellect.

Leech, G. N. (1995), *Semantics*, Harmondsworth: Penguin.

Leech, G., Francis, B. and Xu, X. (1994), 'The Use of Computer Corpora in the Textual Demonstrability of Gradience in Linguistic Categories', in Fuchs, C. and Vitorri, B. (eds), *Continuity in Linguistic Semantics*, Amsterdam and Philadelphia: John Benjamins Publishing Company.

Lees, R. B. (1953), 'The Basis of Glottochronology', *Language* 29, pp. 113–27.

Lehrberger, J. (1982), 'Automatic Translation and the Concept of Sublanguage', in Kettridge, R. and Lehrberger, J. (eds), *Sublanguage: Studies of Language in Restricted Semantic Domains*, Berlin and New York: Walter de Gruyter.

Leighton, J. (1971), 'Sonnets and Computers: An Experiment in Stylistic Analysis Using an Elliott 503 Computer', in Wisbey, R. A. (ed.), *The Computer in Literary and Linguistic Research*, Cambridge: Cambridge University Press.

Loken-Kim, K., Park, Y.-D., Mizunashi, S., Fais, L. and Tomokiyo, M. (1995), 'Verbal-Gestural Behaviour in Multimodal Spoken Language Interpreting Telecommunications', *Eurospeech 95*, vol. 1, pp. 281–4.

Lounsbury, F. G. (1954), 'Transitional Probability, Linguistic Structure and Sytems of Habit-Family Hierarchies', in Osgood, C. E. and Sebeok, T. A. (eds), *Psycholinguistics*, Bloomington: Indiana University Press.

Lovins, J. B. (1969), 'Development of a Stemming Algorithm', *Mechanical Translation and Computational Linguistics*, 11(6), pp. 22–31.

Lowrance, R. and Wagner, R. A. (1975), 'An Extention to the String to String Correction Problem', *Journal of the Association for Computing Machinery*, 22, pp. 177–83.

Lucke, H. (1993), 'Inference of Stochastic Context-Free Grammar Rules From Example Data Using the Theory of Bayesian Belief Propagation', *Proceedings of Eurospeech 93*, pp. 1195–8.

Luk, R. W. P. (1994), 'An IBM Environment for Chinese Corpus Analysis', *Proceedings of COLING*, Kyoto, Japan.

Lutoslawski (1897) (see Kenny 1982).

Lyne, A. A. (1985), *The Vocabulary of French Business Correspondence*, Geneva and Paris: Slatkine-Champion.

Lyne, A. A. (1986), "In Praise of Juilland's D', in *Méthodes Quantitatives et Informatiques dans l'Étude des Textes*, Volume 2, p. 587, Geneva and Paris: Slatkine-Champion.

Lyon, C. and Frank, R. (1992), 'Improving a Speech Interface with a Neural Net', in Beale, R. and Findlay, J. (eds), *Neural Networks and Pattern Recognition in Human Computer Interfaces*, Hemel Hempstead: Ellis Horwood.

McCreight, E. M. (1976), 'A Space-Economical Suffix-Tree Construction Algorithm', *Journal of the Association for Computing Machinery*, 23(2), pp. 262–72.

McEnery, A. M. 'Computational Pragmatics: Probability, Deeming and Uncertain References' (Ph.D. thesis, University of Lancaster, 1995).

McEnery, A., Baker, P. and Wilson, A. (1994), 'The Role of Corpora in Computer Assisted Language Learning', *Computer Assisted Language Learning*, 6 (3), pp. 233–48.

McEnery, A. M. and Oakes, M. P. (1996), 'Sentence and Word Alignment in the CRATER Project', in Thomas, J. and Short, M. (eds), *Using Corpora for Language Research*, London: Longman.

McEnery, A. M. and Oakes, M. P. (1997), 'Authorship Identification and Computational Stylometry', in Dale, R., Moisl, H. and Somers, H. (eds), *Corpus Analysis and Application of Derived Statistics to Natural Language Processing, Part 2: A Handbook of Natural Language Processing*, Basel: Marcel Dekker.

McEnery, A. M. and Wilson, A. (1996), *Corpus Linguistics*, Edinburgh, Edinburgh University Press.

McEnery, A. M., Xu Xin and Piao, S.-L. (1997), 'Parallel Alignment in English and Chinese', in Botley, S., McEnery, A. M. and Wilson, A. (eds), *Multilingual Corpora: Teaching and Research*, Amsterdam: Rodopi.

McKinnon, A. and Webster, R. (1971), 'A Method of "Author" Identification', in Wisbey, R. A. (ed.), *The Computer in Literary and Linguistic Research*, Cambridge: Cambridge University Press.

Maclay, M. and Osgood, C. E. (1959), 'Hesitation Phenomena in Spontaneous English Speech', *Word*, 15, pp. 19–44.

McLeod, W. T. (ed.) (1987), *Collins Dictionary and Thesaurus*, Glasgow: Harper-Collins.

MacQueen, J. B. (1967), 'Some Methods for Classification and Analysis of Multivariate Observations', *Proceedings of the 5th Symposium on Mathematics, Statistics and Probability, Berkeley, 1*, Berkeley: University of California Press, pp. 281–97.

Markov, A. A. (1916), 'Ob odnom primenii statisticheskogo metoda' ('An Application of Statistical Method'), *Izvestiya Imperialisticheskoj akademii nauk*, 6(4), pp. 239–42.

Marshall, I. (1987), 'Tag Selection Using Probabilistic Methods', in Garside, R., Leech, G. and Sampson, G. (eds), *The Computational Analysis of English*, London and New York: Longman.

Mays, E., Damerau, F. J. and Mercer, R. L. (1991), 'Context Based Spelling Correction', *Information Processing and Management*, 27(5), pp. 517–22.

Michaelson, S. and Morton, A. Q. (1976), 'Things Ain't What They Used To Be', in Jones, A. and Churchhouse, R. F. (eds), *The Computer in Literary and Linguistic Studies*, Cardiff: The University of Wales Press.

Milic, L. T. (1966), 'Unconscious Ordering in the Prose of Swift', in Leed, J. (ed.), *The Computer and Literary Style*, Kent, OH: Kent State University Press.

Miller, G. A. (1950), 'Language Engineering', *Journal of the Acoustical Society of America*, 22, pp. 720–5.

Miller, G. and Nicely, P. E. (1955), 'An Analysis of Perceptual Confusion Among English Consonants', *Journal of the Acoustic Society of America*, 27, pp. 338–52.

Milton, J. (1997), 'Exploiting L1 and L2 Corpora in the Design of an Electronic Language Learning and Production Environment', in Grainger, S. (ed.), *Learner English on Computer*, Harlow: Longman.

Mitkov, R. (1996), 'Anaphor Resolution: A Combination of Linguistic and Statistical Approaches', in Botley, S., Glass, J., McEnery, A. and Wilson, A. (eds), 'Approaches to Discourse Anaphora', *Proceedings of the Discourse Anaphora and Anaphora Resolution Colloquium* (DAARC 96), pp. 76–84. Also: UCREL *Technical Paper, Volume 8 (Special Issue)*.

Mitton, R. (1996), 'English Spelling and the Computer', London and New York: Longman.

Mood, A. M., Graybill, F. A. and Boes, D. C. (1974), *Introduction to the Theory of Statistics* (Third Edition, International Student Edition), McGraw-Hill, p. 530.

Morimoto, T. (1995), 'Speaking in Tongues', *Science Spectra*, 1, pp. 33–5.

Morris, R. and Cherry, L. L. (1975), 'Computer Detection of Typographical Errors', IEEE Trans Professional Communication, PC-18 (1), pp. 54–64.

Morrison, D. F. (1990), *Multivariate Statistical Methods*, New York: McGraw-Hill International Editions Statistics Series.

Morton, A. Q. (1965), 'The Authorship of Greek Prose', *Journal of the Royal Statistical Society*, 128(2), pp. 169–233.

Morton, A. Q. (1978), *Literary Detection*, East Grinstead: Bowker Publishing Company.

Mosteller, F. and Wallace, D. L. (1964), *Applied Bayesian and Classical Inference: The Case of the Federalist Papers*, Addison-Wesley.

Nagao, M. (1984), 'A Framework of a Mechanical Translation between Japanese and English by Analogy Principle', in Elithorn, A. and Banerji, R. (eds), *Artificial and Human Intelligence*, Amsterdam: Elsevier Science Publishers.

Nakamura, J. and Sinclair, J. (1995), 'The World of Woman in the Bank of English: Internal Criteria for the Classification of Corpora', *Literary and Linguistic Computing*, 10(2) pp. 99–110.

NJstar (1991–2), Chinese WP System, Version 2.1 Plus. Hongbo Ni, P.O. Box 866, Kensington, NSW 2033, Australia, email Hongbo@csd.unsw.oz.au.

Nothofer, B. (1975), *The Reconstruction of Proto-Malayo-Javanic*, The Hague: Martinus Nijhoff.

Oakes, M. P., 'Automated Assistance in the Formulation of Search Statements for Bibliographic Databases' (Ph.D. thesis, University of Liverpool, 1994).

Oakes, M. P. and Taylor, M. J. (1991), 'Towards an Intelligent Intermediary System for the Ringdoc Drug Literature Database', in Baray, M. and Ozguc, B. (eds), *Proceedings of the Sixth International Symposium on Computer and Information Sciences*, Amsterdam: Elsevier, pp. 51–63.

Oakes, M. P. and Taylor, M. J. (1994), 'Morphological Analysis in Vocabulary Selection for the Ringdoc Pharmacological Database', in Barahona, P., Veloso, M. and Bryant, J. (eds), *Proceedings of the 12th International Congress of the European Federation for Medical Informatics*, pp. 523–8.

Oakes, M. P. and Xu, X. (1994), 'The Production of Concordancing Facilities for the Hua Xia Corpus of Chinese Text', in Moneghan, A. (ed.), *Proceedings of CS-NLP Conference*.

Oakman, R. L. (1980), *Computer Methods for Literary Research*, Columbia: University of South Carolina Press.

Ooi, V. B. Y. (1998), *Computer Corpus Lexicography*, Edinburgh: Edinburgh University Press.

Paducheva, E.V. (1963), 'The Application of Statistical Methods in Linguistic Research', in Akhmanova, O. S. et al. (eds), *Exact Methods in Linguistic Research*, Berkeley and Los Angeles: University of California Press.

Paice, C. D. (1977), 'Information Retrieval and the Computer', *McDonald and Jane's Computer Monographs*, 26.

Paice, C. D. (1990), 'Another Stemmer', *SIGIR Forum* 24, pp. 56–61.

Paice, C. D. (1996), 'Method for Evaluation of Stemming Algorithms Based on Error Counting', *Journal of the American Society for Information Science* (JASIS), 47(8), pp. 632–49.

Parrish, S. M. (1959), *A Concordance to the Poems of Matthew Arnold*, Ithaca: Cornell University Press.

Pei, M. A. and Gaynor, F. (1954), *A Dictionary of Linguistics*, London: Peter Owen.

Phillips, M. (1989), 'Lexical Structure of Text', *Discourse Analysis Monograph 12*, English Language Research, University of Birmingham.

Pollock, J. J. and Zamora, A. (1984), 'Automatic Spelling Correction in Scientific and Scholarly Text', *Communications of the ACM*, 27(4), pp. 358–68.

Porter, M. F. (1980), 'An Algorithm for Suffix Stripping', *Program*, 14, p. 130.

Pratt, F. (1942), *Secret and Urgent*, Garden City, NY: Blue Ribbon Books.

Rabiner, L. R. (1989), 'A Tutorial on Hidden Markov Models and Selected Applications in Speech Recognition', *Proceedings of the IEEE*, 77(2), pp. 257–86.

Renouf, A. and Sinclair, J. (1991), 'Collocational Frameworks in English', in Aijmer, K. and Altenberg, B. (eds), *English Corpus Linguistics*, London and New York: Longman.

Riseman, E. M. and Hanson, A. R. (1974), 'A Contextual Post-Processing System for Error Correction Using Binary N-grams', *IEEE Trans Computers C-23* (5), pp. 480–93.

Ritter (1888) (see Kenny 1982).

Robertson, A. M. and Willett, P. (1992), 'Evaluation of Techniques for the Conflation of Modern and Seventeenth Century English Spelling', in McEnery, A. and Paice, C. (eds), *Proceedings of the 14th Information Retrieval Colloquium*, Lancaster, pp. 155–65.

Rosengren, I. (1971), 'The Quantitative Concept of Language and its Relation to the Structure of Frequency Dictionaries', Études de Linguistique Appliquée (Nlle Sér.), 1: pp. 103–27.

Ross, D. (1973), 'Beyond the Concordance: Algorithms for Description of English Clauses and Phrases', in Aitken, A. J., Bailey, R. W. and Hamilton-Smith, N. (eds), *The Computer and Literary Studies*, Edinburgh: Edinburgh University Press.

Rudall, B. H. and Corns, T. N. (1987), *Computers and Literature*, Cambridge, MA and Tunbridge Wells, Kent: Abacus Press.

Salton, G. and McGill, M. J. (1983), *Introduction to Modern Information Retrieval*, New York: McGraw-Hill.

Sampson, G. (1987), 'Probabilistic Models of Analysis', in Garside, R., Leech, G. and Sampson, G. (eds), *The Computational Analysis of English*, London and New York: Longman.

Sapir, E. (1921), *Language*, Harcourt, Brace & World.

Savoy, J. (1993), 'Stemming of French Words Based on Grammatical Categories', JASIS 44(1), pp. 1–9.

Schinke, R., Greengrass, M., Robertson, A. M. and Willett, P. (1996), 'A Stemming Algorithm for Latin Text Databases', *Journal of Documentation*, 52(2), pp. 172–87.

Schütz, A. J. and Wenker, J. (1966), 'A Program for the Determination of Lexical Similarity Between Dialects', in Garvin, P. L. and Spolsky, B. (eds), *Computation in Linguistics*, Bloomington and London: Indiana University Press.

Scott, Mike (1996), *WordSmith Tools Manual*, Oxford: Oxford University Press.

Sekine, S. (1994), 'A New Direction for Sublanguage NLP', in Jones, D. (ed.), *Proceedings of the International Conference on New Methods in Language Processing* (NeMLaP), UMIST, Manchester, pp. 123–9.

Sermon, R. (1996a), 'The Hackness Cross Cryptic Inscriptions', Gloucester Archaeology.

Sermon, R. (1996b), 'The Hackness Cross Cryptic Inscriptions', *Yorkshire Archaeological Journal*, 68, pp. 101–11.

Shannon, C. E. (1949), 'The Mathematical Theory of Communication', in Shannon, C. and Weaver, W. (eds), *The Mathematical Theory of Communication*, Urbana, IL: The University of Illinois Press.

Sharman, R. A. (1989), 'An Introduction to the Theory of Language Models', IBM UKSC Report 204, July 1989.

Shepard, R. N. (1972), 'Psychological Representation of Speech Sounds', in David, E. E. and Denes, P. B. (eds), *Human Communication: A Unified View*, New York: McGraw-Hill.

Sherman (1888) (see Kenny 1982).

Shtrikman, S. (1994), 'Some Comments on Zipf's Law for the Chinese Language', *Journal of Information Science*, 20(2), pp. 142–3.

Simard, M., Foster, G. and Isabelle, P. (1992), 'Using Cognates to Align Sentences in Bilingual Corpora', *Proceedings of the Fourth International Conference on Theoretical and Methodological Issues in Machine Translation (TMI92)*, Montreal, Canada, pp. 67–81.

Sklar, B. (1988), *Digital Communications: Fundamentals and Applications*, Englewood Cliffs, NJ: Prentice-Hall.

Smadja, F. (1991), 'Macro-coding the Lexicon with Co-Occurrence Knowledge', in Zernik, U. (ed.), *Lexical Acquisition: Exploiting On-Line Resources to Build a Lexicon*, Hillsdale, NJ: Lawrence Erlbaum Associates.

Smajda, F. (1992), 'XTRACT: An Overview', *Computers and the Humanities*, 26(5–6), pp. 399–414.

Smadja, F., McKeown, K. R. and Hatzivassiloglou, V. (1996), 'Translating Collocations for Bilingual Lexicons: A Statistical Approach', *Computational Linguistics*, 22(1), pp. 1–38.

Small, H. and Sweeney, E. (1985), 'Clustering the Science Citation Index using Cocitations', *Scientometrics*, 7, pp. 391–409.

Sneath, P. H. A. and Sokal, R. R. (1973), *Numerical Taxonomy*, San Francisco: W. H. Freeman and Co.

Somers, H. H. (1961), 'The Measurement of Grammatical Constraints', *Language and Speech*, 4, pp. 150–6.

Spearman, C. (1904), 'The Proof and Measurement of Association Between Two Things', *American Journal of Psychology*, 15, pp. 88–103.

Sperber, D. and Wilson, D. (1986), 'Relevance: Communication and Cognition', Oxford: Blackwell.

Svartvik, J. (1966), *On Voice in the English Verb*, The Hague: Mouton.

Swadesh, M. (1952), 'Lexicostatistical Dating of Prehistoric Ethnic Contacts', *Proceedings of the American Philosophical Society 96*, pp. 453–63.

Tannenbaum, P. H., Williams, F. and Hillier, C. S. (1965), 'Word Predictability in the Environment of Hesitations', *Journal of Verbal Learning and Verbal Behaviour*, 4, pp. 118–28.

Thisted, R. and Efron, B. (1987), 'Did Shakespeare Write a Newly-Discovered Poem?', *Biometrika*, 74(3), pp. 445–55.

Thompson, W. and Thompson, B. (1991), 'Overturning the Category Bucket', *Byte*, 16(1), pp. 249–56.

Tottie, G. (1991), *Negation in English Speech and Writing*, San Diego, Academic Press.

Tweedie, F. J., Singh, S. and Holmes, D. I. (1994), 'Neural Network Applications in Stylometry: The Federalist Papers', in Moneghan, A. (ed.), *Proceedings of CS-NLP 1994*, Natural Language Group, Dublin City University.

van Mises, R. (1931), *Wahrscheinlichkeitsrechnung und Ihre Anwendung in der Statistik und Theoretischen Physik*, Leipzig and Vienna, pp. 415–18.

van Rijsbergen, C. J. (1979), *Information Retrieval*, Butterworths, London.

Wagner, H. (1958), *Linguistic Atlas and Survey of Irish Dialects*, Dublin: Dublin Institute for Advanced Studies, 1958–69, 4 vols.

Wagner, R. A. and Fischer, M. J. (1974), 'The String to String Correction Problem', *Journal of the Association for Computing Machinery*, 21, p. 168.

Walker, S. (1988), 'Improving Subject Access Painlessly: Recent Work on the OKAPI Online Catalogue Projects', *Program*, 22(1), pp. 21–31.

Ward, J. H. (1963), 'Hierarchical Grouping to Optimize an Objective Function', *Journal of the American Statistical Asssociation*, 58(301), pp. 236–44.

Weaver, W. (1949), 'Recent Contributions to the Mathematical Theory of Communication', in Shannon, C. E. and Weaver, W. (eds), *The Mathematical Theory of Communication*, Urbana: The University of Illinois Press.

Willett, P. (1980), 'Document Clustering Using an Inverted File Approach', *Journal of*

Information Science, 2(5), pp. 223–31.

Willett, P. (1988), 'Recent Trends in Hierarchic Document Clustering: A Critical Review', *Information Processing and Management*, 24(5), pp. 577–97.

Williams, C. B. (1970), *Style and Vocabulary*, London: Griffin.

Williams, W.T. and Clifford, H.T. (1971), 'On the Comparison of Two Classifications of the Same Set of Elements', *Taxon*, 20, pp. 519–22.

Wisbey, R. (1971), 'Publications from an Archive of Computer-Readable Literary Texts, in Wisbey, R. (ed.), *The Computer in Literary and Linguistic Research*, Cambridge: Cambridge University Press.

Woods, A., Fletcher, P. and Hughes, A. (1986), *Statistics in Language Studies*, Cambridge: Cambridge University Press.

Wolff, J. G. (1991), *Towards a Theory of Cognition and Computing*, Chichester: Ellis Horwood.

Xu Xin (1996), 'Building and Aligning an English-Chinese Parallel Corpus' (MA Dissertation, Department of Linguistics and MEL, University of Lancaster).

Yamamoto, M. (1996), 'The Concept of Animacy and its Reflection in English and Japanese Referential Expressions' (Ph.D. thesis, Department of Linguistics and MEL, University of Lancaster).

Yannakoudakis, E. J. and Angelidakis, G. (1986), 'An Insight into the Entropy and Redundancy of the English Dictionary', *IEEE Transactions on Pattern Analysis and Machine Intelligence*, 10(6), pp. 960–70.

Yannakoudakis, E. J. and Fawthrop, D. (1983a), 'The Rules of Spelling Errors', *Information Processing and Management*, 19(2), pp. 87–99.

Yannakoudakis, E. J. and Fawthrop, D. (1983b), 'An Intelligent Spelling Error Corrector', *Information Processing and Management*, 19(2), pp. 101–8.

Yardi, M. R. (1946), 'A Statistical Approach to the Problem of the Chronology of Shakespeare's Plays', *Sankhya* (Indian Journal of Statistics), 7(3), pp. 263–8.

Yule, G. U. (1939), 'On Sentence Length as a Statistical Characteristic of Style in Prose, with Applications to Two Cases of Disputed Authorship', *Biometrika* 30, 363–90.

Yule, G. U. (1944), 'The Statistical Study of Literary Vocabulary', Cambridge: Cambridge University Press.

Zernik, U. (1991), 'Train 1 vs Train 2: Tagging Word Sense in a Corpus', in Zernik, U. (ed.), *Lexical Acquisition: Exploiting On-line Resources to Build a Lexicon*, Hillsdale, NJ: Lawrence Erlbaum Associates.

Zipf, G. K. (1935), *The Psycho-Biology of Language*, Boston, MA: Houghton Mifflin.

Zylinski (1906) (see Kenny 1982).

Index